T0146207

Making Computers Accessible

Making Computers Accessible

Disability Rights and Digital Technology

Elizabeth R. Petrick

Johns Hopkins University Press
Baltimore

Johns Hopkins University Press
2715 North Charles Street
Baltimore, Maryland 21218-4363
www.press.jhu.edu

Library of Congress Cataloging-in-Publication Data
Petrick, Elizabeth, 1978–
   Making computers accessible : disability rights and digital technology /
Elizabeth Petrick.
       pages cm. — (History of science, technology, and medicine)
   Includes bibliographical references and index.
ISBN 978-1-4214-1646-5 (hardcover : alk. paper) — ISBN 978-1-4214-1647-2
(electronic) — ISBN 1-4214-1646-8 (hardcover : alk. paper) —
ISBN 1-4214-1647-6 (electronic)   1. Computers and people with
disabilities—United States.   2. Assistive computer technology—
United States.   3. Microcomputers—Social aspects—
United States.   I. Title.
   HV1569.5.P47 2015
   004.16087—dc23   2014033485

A catalog record for this book is available from the British Library.

*Special discounts are available for bulk purchases of this book. For more
information, please contact Special Sales at 410-516-6936 or
specialsales@press.jhu.edu.*

Johns Hopkins University Press uses environmentally friendly book materials,
including recycled text paper that is composed of at least 30 percent
post-consumer waste, whenever possible.

Contents

Acknowledgments

This book would not have been possible without the help and guidance of many people. I began this project as a PhD student in History and Science Studies at the University of California, San Diego. I would like to thank my graduate advisor, Cathy Gere, who did her best to make me a better writer, and I will always appreciate it. I also want to thank Charles Thorpe for making the time to run a dissertation workshop group with all of his students; his and their feedback was invaluable, particularly in helping me to consider interdisciplinary views. I am grateful to Tal Golan, Robert Westman, and Kelly Gates for their encouragement and feedback along the way. The graduate students also helped me, throughout my time there, by letting me bounce ideas off of them and providing necessary critique, in particular: Kevin Walsh, Emily Lee, Monica Hoffman, Katrina Petersen, Krystal Tribbett-McCants, Bob Long, Leah Cluff, and Jon Stern.

I would also like to thank some of the people this book is about, whose history I have told, especially the Brand family and Alan Brightman. Their work to promote accessible computer technologies for people with disabilities made this book possible and also gave me an often intimate look at the role of individual people in the technology development process. The materials on their work and that of others mostly came from oral history interviews they conducted and from the archive at the Bancroft Library of the University of California, Berkeley. In addition, Alan met and talked with me, offering me much-needed insights into the accessibility work done at Apple Computer. Jackie Brand provided me with significant materials from her personal collection, both photographs and documents, and encouraged my telling of this history. I thank her daughter, Shoshana Brand, for allowing me to share images of her own use of accessible personal computer technology and thus put a face to the significance of this technology.

I also appreciate Robert J. Brugger at Johns Hopkins University Press for his patience and assistance in guiding me through this publication process. I am grateful as well for the feedback in the review done by Howard Segal from the University of Maine.

I cannot thank my friends and family enough for their support in getting me here and enabling me to complete this book. In particular, I want to thank Holly Zynda for her copyediting expertise, Kristofer Lee for endless conversations in which I mulled over problems, and Matt McGuire for providing me with operating system images. Finally, I never could have done this without the support of my parents, who encouraged me to change disciplines entirely, from computer science to history, and work toward a goal that seemed a very long way off when I started.

Making Computers Accessible

The Development of Accessible Personal Computer Technologies

With regard to the major physical and sensory disabilities, I believe that in a couple of decades we will come to herald the effective end of handicaps. As amplifiers of human thought, computers have great potential to assist human expression and to expand creativity for all of us.

RAYMOND KURZWEIL, *THE AGE OF SPIRITUAL MACHINES: WHEN COMPUTERS EXCEED HUMAN INTELLIGENCE*

In 1974, not long after developing the first optical character recognition technology that could recognize any printed text, Raymond Kurzweil sat next to a blind man on a plane flight. Kurzweil was searching for a use for his new software—looking for a problem to which the software might provide the solution. At this time, no computer program existed which could translate text of any font into speech for blind people to understand. The man on the flight described this limitation as "the only real handicap" his blindness presented. Kurzweil, a computer technology developer and later futurist philosopher, decided that this would be the use to which he would put his new innovation, to "overcome this principal handicap of blindness," as he saw it—that is, the inability to read printed texts.[1] By 1976, the Kurzweil Reading Machine existed as a working prototype that could scan printed texts and translate them into speech. People expressed immediate interest in the technology; after seeing it featured on the *Today Show*, Stevie Wonder contacted Kurzweil's company and was given their second-ever version of the machine.[2] The Kurzweil Reading Machine would set the bar for other text-to-speech software in terms of accuracy and vocabulary. The Reading Machine was not Kurzweil's only accessible technology innovation. In 1982, his company created the Kurzweil Voice, general purpose dictation software that was the descendent of specialized medical dictation technology. Kurzweil found that this software was particularly useful for people whose disability affected the use of their hands; it allowed them to compose text in word-processing software by speaking.[3] These early accessible computer technologies placed Kurzweil at the forefront of innovations for people with disabilities and demonstrated the possibilities of computers to improve people's lives.

Looking toward the future, Kurzweil predicted that one day advanced computer technologies would eradicate disability itself, first by accommodating all the needs of people with disabilities and later by fixing their bodies. In Kurzweil's vision of the future, computer technology created for people with disabilities acted as the first step in people transcending the limitations of their bodies, thus altering what it means to be human. This transhumanist utopian vision descends from utopian ideals dating back to Plato and defined by Thomas More's 1516 novel, *Utopia*. Kurzweil's utopia follows from Enlightenment dreams that flowed into the Industrial Revolution and looked not to an ideal society hidden away, waiting to be found, but to the future, when science and technology would inevitably improve society to a point where utopia might be attainable.[4] For most utopianists, the Christian belief in original sin prevented the full realization of any true utopian society for human beings; they could only be improved upon, not made perfect. Unlike this idea of an inherently limited utopia, Kurzweil's vision of a better future is one where all problems with the human body are fixed by technology, and, ultimately, the body itself is left behind. Technology is capable of solving all human physical problems, eliminating the barriers that prevent certain people from fully participating in society but also offering the potential to transform human existence itself. Technologists, Kurzweil believed, could aid the creation of this future by working on innovations that aided people in the present.

Disability and technology activists also looked at developing computer technology and saw its potential to overcome barriers and improve the lives of people with disabilities, taking a celebratory perspective toward the use of computers. Unlike Kurzweil, however, these activists did not reach toward a utopia in some distant future. Instead they focused on the daily reality of people with disabilities struggling for fuller participation in society. Theirs was a more pragmatic vision of technological potential, promoting the development of computer technologies, such as those Kurzweil himself created, but with an emphasis on the unique needs of individuals and the often messy realities of using technology to augment their abilities. Disability activists focused on people with disabilities as technology users, as people who might incorporate technology into their lives in order to meet their individual needs and desires.

Throughout the history of accessible computer technologies, people with disabilities acted as the paradigmatic computer users. Technology created for their use was the foundation for technologies intended to augment all humans, a goal in line with Kurzweil's arguments for technology benefiting people. Developers had to design specifically for people with disabilities in mind—making their needs primary—before they could expand the scope of computer technology to include everyone. This was not a one-way relationship between developers and users, however; people with disabilities and nondisabled activists pushed for changes to personal computers that would make them more inclusive of different needs. This history of the creation of accessible personal computer technologies did not play out in a futurist vision of ever-accelerating technological progress toward a state of bodilessness but rather in the actual development of personal computer technologies for people with disabilities. Innovators, users, activists, and policy makers worked to build an inclusive technology instead of one developed according to ideas of normalcy that excluded people with disabilities, to put their needs at the forefront of the development process.

The role of everyday users, and people with disabilities in particular, lies mostly unexplored within the history of computers. This history tells us about the influences of the military and counterculture in creating computers and embedding them with values that influenced the place of the technology in society, as well as about the importance of innovators, companies, and computer professionals. Hackers or people who blur the line between users and developers, such as hobbyists or computer professionals, generally define the "user." We know less about how everyday consumers play a part in the development of computers or how the design of technology affects people's use of it. From the perspective of people with disabilities, various rich histories detail the civil rights struggle and the use of assistive technologies, but fewer studies examine everyday technologies. The history of accessible personal computer technologies for people with disabilities combines the history of computers and the history of people with disabilities in order to understand the role of users in the development of computers and the significance of computers in the lives of users.

By accommodating physical needs, personal computer technology played a political role in the historical enactment of civil rights for people with disabilities. The development of accessible technologies intersected with the history of civil rights and the emergence of identity politics; as disability became an identity that people could claim and as their consciousness as a marginalized population within society grew, people with disabilities increasingly advocated for changes to the social environment that would address concerns about equity. Technological accommodation, in the form of personal technology and in the public built environment, is necessary for people with disabilities to experience full participation in a society that has not been built with their needs in mind. Technology alone, however, does not change anyone's lives for the better; people must be aware of it, and it must be available to them for its benefits to be realized. This element of technological development and use does not appear in Kurzweil's account of technology and society. Activists created social technologies, in the form of networked advocacy groups, to share information and disseminate knowledge to consumers. In some instances, federal antidiscrimination legislation was necessary to mandate access to computer technologies.

Accessible personal computer technology took three forms during this history: a political technology, which made equity possible; a legal technology, which was required and funded by the federal government; and a social technology, which created new forms of information sharing and communication. For its potential as a political technology to be realized, developers needed to build these technologies into the personal computer, and corporate philanthropy played an essential role in creating accessible computer technologies and supplying them to users. First though, developers had to rethink who counted as a user in order for these technologies to work for a variety of uses. By embracing the values behind what would become universal design, developers met the needs of different people by creating technologies that were flexible enough to accommodate bodily variations. This approach started with hardware and the physical input and output devices people use to interact with the computer, then it was applied to creating software that allowed people to experience the same computer functions in different ways. The federal government mandated some of these accessibility developments in order for the civil rights potential of the technology to be realized. As a social technology, the personal computer was a part of its own network of technological dissemination—informing people of what was available and what was possible. On a more personal level, the technology also enabled new forms of communication for users. These three aspects of the development of accessible personal computer technology fit in well with Kurzweil's philosophy of technology changing individual lives, but beyond the personal lay a social level absent from his view, where technology was built within companies, promoted and disseminated by activists, talked about and requested by users, and legislated by the government.

All three aspects of the technology—political, legal, and social—came together in the form of technological accommodations to enact civil rights. For these rights to be realized, access had to be built into the technology. The hard, messy process of creating accessible technologies is the opposite of Kurzweil's valorized posthuman account of the history of technology; the development of accessibility was fundamentally *human*, a part of political movements with a long history of struggling for equality. This history was practical, not visionary, a search for solutions to individual and social problems. Technology such as braille printed on signs in buildings or closed captioning on TV can provide a means to physically overcome some of the barriers in society that disable people, but people can also apply technology more problematically, as a way to "fix" the bodies of people with disabilities, treating people rather than their environment as the problem that needs solving. The activists and developers in this history tended to view technology in utopian and celebratory terms, applauding its potential as a means to encourage future development. Yet new personal computer technologies also introduced new barriers and pushed the argument that users had to adapt to the technology; technology need not change to accommodate them. Today, the personal computer has come to fulfill some of its early promise, but it has hardly done so in a straightforward fashion. The relationship between the technology and those who used and developed it involved incessant labor on the part of people with disabilities and activists. They worked to communicate the needs of people with different bodies and influence the direction of a developing technology. They also instituted standards that provided access instead of introducing barriers.

Researchers and innovators, before the personal computer, described computers as a technology of human augmentation—one that could provide new kinds of physical and intellectual capabilities. People with disabilities particularly benefited from the possibilities computers offered. From the perspective of the computer, as imagined by its designers, all people possess limitations that the computer could accommodate. People with disabilities stood as the model users, the fundamental use case for computer technology. Efforts focused on meeting their needs coincided with a new paradigm for personal computers of user-friendliness and individual, everyday use. This emphasis on usability overlapped with the gradual emergence of ideals of universal design. Universal design called for a consideration of the needs of all users in order for a technology to be usable by as many people as possible. Instead of designing technology for the "average" or "normal" user—the universal human being—universal design advocated a new understanding of universality: a technology is universal when it can accommodate all differences in how people need to use it. By focusing attention on the seemingly hard cases—the specialized needs of people with different sensory and physical abilities—a technology becomes more flexible and usable. Computer companies slowly realized the benefits of designing their technology to accommodate more users, as doing so expanded their market share to include people with disabilities as consumers and allowed consumers more options in how they chose to use the technology. This approach to design marked a gradual shift in general corporate philosophy from a philanthropic view of people with disabilities to one in which they were a viable market share.

Personal computers differ from specialized, assistive technologies because they are a general, consumer technology. They provide for many different kinds of uses, which affect people's abilities, but these uses focus more on different ways of using the technology than specifically on types of disability. Like buildings, personal computers had to be made accessible, with features built into them that allowed people with a variety of bodies and abilities to use them. The individualized nature of the technology means that personal computer accessibility may have different meanings for people with different disabilities and affect them in different ways. For some people with disabilities, an everyday technology needed to be accessible so that it could be used for the same purposes desired by people without disabilities. For others, however, the personal computer was itself an assistive device, granting abilities, such as methods of communication, a person might otherwise not have. The computer carried with it—embedded into its development over time—values of being a universalizing technology with the potential to be used by anyone for any purpose, values similar to those promoted by universal design.

To capture the history of this multiplicity of uses, the terms *personal computer* and *people with disabilities* operate in this book as broad, umbrella categories. *Personal computer* includes any technologies related to personal computers that people with disabilities might use, and *people with disabilities* refers to those with any type of impairment that affects their use of personal computer technology. There exists no current consensus on whether to use the phrase *people with disabilities* or *disabled people*. Proponents of both sides make strong arguments for why one is preferable over the other. Currently, U.S. academia chooses to use *people with disabilities,* as it places the emphasis on the people first and represents disability as secondary. Groups with certain disabilities, such as blind people and deaf people, prefer to be referred to as such, and I utilize their chosen descriptors. Terminology preferences have changed throughout the past few decades, and the actors in this history often use terms considered offensive today. In addition to the issue of labels, it is useful to remember that categories of disability are not absolute and have no fixed boundaries. As to who falls within *people with disabilities*, I use the broad definition found in U.S. laws such as the Americans with Disabilities Act, which includes anyone who has some impairment that affects a major life activity. The activist groups I study tend to be as inclusive as possible, therefore I also err on the side of inclusivity.

Three terms appear throughout this history which have some degree of overlap and are to some extent interchangeable. The broadest phrase is *accessible technologies*, which refers to those devices, software, or features intended to provide access to a technology or ability. It frequently pertains to people with disabilities, but it can also include the access needs of other marginalized groups. For example, television closed captioning or screen-reading software that translates text to speech are accessible technologies. *Assistive technologies* are specifically intended to aid people with disabilities in some way, such as wheelchairs or computer keyboards with large, easy-to-press keys. *Adaptive devices* allow access based on a specific disability. Many of the technologies discussed in this book fit under all three of these terms, but for the most part, I utilize *accessible technologies* when discussing personal computer devices and features.

The development of these accessible technologies began with pre–personal computer technologies for people with disabilities and civil rights legislation during the 1970s. Once the personal computer came on the consumer market in the late 1970s, early innovations started to make them accessible for people with disabilities. The technology stabilized during the mid-1980s, as developers set standards. The mid-1990s saw a new kind of user interface, and by the late 1990s, accessibility become nearly mainstream within computer development; it was a part of the framework of technological design and was followed as a design goal, to varying degrees of success.

Computer companies—mainly Apple Computer and IBM—and disability activists, such as the Disabled Children's Computer Group (DCCG) and the Alliance for Technology Access (ATA), were strongly intertwined throughout their history. Located in the Bay Area, Apple Computer and the DCCG came together in the mid-1980s to create the ATA, which acted as an umbrella organization to combine the efforts of dozens of small groups across the country. The ATA began its work from within Apple before separating from the company to become an independent nonprofit disability and technology advocacy group. Apple continued its ties to these groups into the 1990s, and IBM also played an integral role in providing funding and technological resources to the ATA. Accessible personal computer technologies developed through negotiation and feedback among developers, activists, and users. These people came together in an attempt to fulfill the promise that the technology holds for enacting civil rights for people with disabilities and for creating more usable, flexible technologies that provide people with new kinds of abilities and sites of social interaction.

Chapter One

Disability Rights and Technology before the Personal Computer

Prior to the invention of the personal computer, there existed two significant arenas for the intersection of people with disabilities and computer technology. The first was disability rights legislation, which, during the 1960s and 1970s, started to offer people with disabilities the promise of equal rights, enacted through technological accommodations. The second was the use of early computer technology by people with disabilities, both professionally and as research subjects. The histories of these two arenas of intersection came together in the form of the personal computer, a technology that could provide users with new abilities and grant access to new forms of social participation. In analyzing these two arenas, I introduce five historical themes: (1) the shift in disability rights from a paternalistic, caretaker model to a civil rights–based model; (2) the concomitant need for technological accommodations in order for people with disabilities to achieve equal rights; (3) the idea of the computer as a universalizing technology, a tool for anyone, for any purpose; (4) the view of the computer as a technology of augmentation which allows people to expand their cognitive abilities beyond the limits imposed by the body; and (5) the lived reality of different practices of using a computer by people with different bodies.

Using these five themes, I explore a number of significant pieces of legislation and court challenges regarding disability rights. I begin with the history of the rights of people with disabilities in the late nineteenth and early twentieth centuries, prior to civil rights legislation. I trace the development of disability rights through notable federal laws that established enforceable rights, from the Architectural Barriers Act of 1968, to the Rehabilitation Act of 1973 and the creation of its regulations, and finally to the Education for All Handicapped Children Act of 1975.[1] Next, I examine disability activism and accessible personal computer technologies from the perspective of the computer industry and professional organizations. I discuss the situation of blind computer programmers during the 1970s and the technological accommodations that allowed them to perform their jobs. I also examine a professional organization that focused on promoting the employment of disabled computer professionals and disseminating information on technologies that would benefit people with disabilities. Then I look at one of the technologies this organization helped to promote—computerized conferencing—and the research into it targeting people with disabilities. Computerized conferencing demonstrates the potential effect personal computer technology can have on users with disabilities as well as the role in development of considering people with disabilities as intended users. Finally, I look more directly at the computer industry to discuss accessibility work and the training of people with disabilities done by IBM prior to the advent of the personal computer.

Disability Rights Legislation and Enforcement

The history of disability rights shows a transition from a caretaker model, in which society looked after those unable to care for themselves, to a model of fighting for the acquisition of civil rights and equal participation in society through technological accommodations. Prior to civil rights legislation in the 1960s and 1970s, disability legislation addressed the setting up and funding of programs and services, such as vocational rehabilitation, federally funded institutions and schools, and medical care. These laws grew out of the then-common approaches to dealing with people with disabilities by either making people seen as useful into contributors to society or taking care of those seen as unable to contribute. From the late nineteenth century and into the Progressive Era, people with disabilities were considered members of the "deserving poor," dependent on society for their well-being as a result of circumstances beyond their control.[2] As a whole, U.S. society did not view people with disabilities as capable of independence and self-sufficiency; they were a burden that able-bodied society should carry through the work of charities. There was little federal involvement in programs for disabled people during the last half of the nineteenth century. In 1854, President Franklin Pierce vetoed the first bill to pass Congress that would have provided federal funding for institutions for the deaf, dumb, blind, and mentally ill. The president's veto set a precedent of federal government staying out of disability concerns until the twentieth century.[3]

With notable exceptions, there was a lack of federal involvement in addressing problems faced by people with disabilities in the early twentieth century. It wasn't until the 1930s that the federal government created a basic social safety net in response to the Great Depression, but people with disabilities were not its major concern, for a number of reasons. First, belief in the ideals of individualism and upward mobility through hard work and perseverance meant that many Americans were reluctant to see the federal government become involved in social welfare; excessive reliance on government aid to solve social problems was a danger to the strength of the American character. At the same time, however, the widespread view of people with disabilities as members of the deserving poor labeled them an appropriate beneficiary of religious and private charity. Private charity offered a partial solution to the problem of needing to care for people while avoiding the socialist connotations of government aid programs.

The legislation that lawmakers did pass in the first half of the twentieth century responded to three factors that enlarged the disabled population: war, industry, and medical advances. Disability legislation began to focus on how to turn people with disabilities into productive members of society. Vocational rehabilitation programs in the early twentieth century handled the growing disabled population created by the World Wars by providing services for disabled veterans to enter the workplace. The Veterans Bureau began between the wars, with the first Veterans Administration hospitals opening in 1922. These programs neatly resolved the challenge of showing gratitude to veterans for military service, but, unlike previous methods of direct compensation, they did so in a way that also served the nation's welfare by creating workers who could contribute to society.[4] Industrial accidents, combined with improvements in medicine that allowed people with disabilities to live longer, also contributed to a growing disabled population. In 1920, the Smith-Fess Act extended vocational training programs to include all disabled civilians who could potentially work.[5] Vocational rehabilitation programs continued to expand through the mid-twentieth century, increasing their resources and encompassing more people. Although people with disabilities were seen as dependent on these vocational rehabilitation services, they were also seen as capable of easing the burden they placed on society by contributing to the workforce. However, there were no requirements that employers treat employees with disabilities equally or that they provide accommodations for them. Disability legislation and programs prior to World War II mostly came about as a result of governmental or charitable group action. After World War II, the origins of the later disability rights movement and activist groups arose among parents of children with disabilities. As medical advances allowed children to survive disabilities that had previously been fatal, a growing parents' movement sought support and protection through the creation of groups such as the United Cerebral Palsy Association, formed in 1948, and the Muscular Dystrophy Association, formed in 1950.[6]

It was not until the second half of the twentieth century that activists fought to protect the civil rights of Americans with disabilities by calling for technological accommodations that would allow them to participate in society. These activists' arguments echoed those made almost a century earlier in the 1870 case *Sleeper v. Sandown*. David T. Sleeper sued the town of Sandown, New Hampshire, for failing to maintain a bridge railing, which caused him to fall off the bridge. Sleeper accused the town of negligence, and the jury sided with him. The town appealed to the state Supreme Court, arguing that Sleeper was the negligent party because he had been traveling alone in spite of the fact that he was blind. The court again sided with Sleeper, concluding that while a blind person must be more cautious than a sighted person, he had the same right as anyone to walk along the roadway and to assume a bridge railing would be in good repair.[7] The town of Sandown was responsible for maintaining the safety railing that would allow a blind person to use the public roadway independently. In this case, the court ruled that society was obligated to provide accommodations that would allow people with disabilities to move about in the same fashion as anyone else. This was only a local decision, however, as there existed no legislative framework to require accommodations for people with disabilities or to guarantee their participation in society. It was not until almost a century later that civil rights for people with disabilities became a federal issue.

A marked shift in disability rights legislation began in the 1960s with the expansion of the idea of civil rights to include more and more groups of people. The emergence of identity politics brought people together into communities based on social categories of race, gender, sexuality, and eventually, disability, giving new strength to struggles for equality. Landmark bills in disability legislation began to grant equal rights and antidiscrimination protections to people with disabilities. Programs of integration required the inclusion of people with disabilities alongside everyone else—in education, employment, and public services.[8] As disability legislation moved to a rights-based model in the late 1960s, technologies of accommodation became the means through which to enact enforcement. In order to acquire equal rights, people with disabilities needed social spaces to be physically accessible, so buildings and public transportation had to be modified to allow all people access. In the 1970s, prohibitions against discrimination in education and employment made special accommodations for people with disabilities a requirement for programs receiving federal funding.

THE ARCHITECTURAL BARRIERS ACT OF 1968

The Architectural Barriers Act of 1968 (ABA), one of the first pieces of federal disability rights legislation to protect all people with disabilities, mandated technological changes to the built environment to accommodate the needs of people. The purpose of the ABA was "to insure that certain buildings financed with Federal funds are so designed and constructed as to be accessible to the physically disabled."[9] It covered all buildings constructed, leased, or financed by the federal government after August 12, 1968. The act itself set no standards for how accessibility was to be achieved, instead allocating the responsibility for different kinds of buildings to their respective federal agencies—the Department of Housing and Urban Development, the Defense Department, and the Postal Service—all under the oversight of the Department of Health, Education, and Welfare.

Vocational rehabilitation programs supported the move toward barrier-free architecture, as providing training for employment meant little if the buildings that housed jobs were inaccessible.[10] Kent Hull, author of a 1979 American Civil Liberties Union (ACLU) guide to civil rights for people with disabilities, succinctly describes the importance of the problem of architectural barriers: "For many handicapped persons—not just those with ambulatory handicaps, but also blind persons and deaf persons (who face barriers when appropriate stimuli such as brailled elevator buttons and visual public announcement systems are absent)—the existence of architectural barriers is a fact that cannot be discarded by public declarations in favor of equality for handicapped persons."[11] The physical reality of architectural barriers, in other words, makes participation in society a continuous difficulty for people with disabilities. The ABA demonstrates the shift that began in the 1960s when equal participation in society started to be seen as a right of people with disabilities. However, removing architectural barriers had to be explicitly dealt with through enforceable legislation; attempts to encourage the public to remove barriers through education and awareness of the problems they caused had failed.[12] Yet, even federal legislation could not assure compliance when it faced economic concerns. Unlike for other groups of people, antidiscrimination for people with disabilities cannot be enacted through legislation alone. The lack of accessibility in public spaces can only be remedied with significant investment in infrastructural changes, and sometimes the financial costs proved to be an obstacle.

As an example of the kinds of battles fought over the requirement of accessibility in public buildings, in 1970, lawmakers amended the ABA to include the Washington, D.C., Metro system then under construction. Accessibility to the Metro had to be enforced through a 1972 lawsuit brought against the Washington Metropolitan Area Transit Authority by a civil rights organization, the Washington Urban League. The suit focused on issues of accessibility at the Gallery Place station. As the Metro became operational in 1976, Gallery Place station was still closed due to lack of accessibility. Local businesses requested that the federal court order the opening of the station, as they were losing customers to businesses near functional stations. The court refused, appearing to choose to defend the rights of people with disabilities and enforce existing legislation over the economic concerns of local business; the judge explained the necessity of forced compliance with the ABA.

> There is simply no other way apparent to the Court to ensure that the defendant, once and for all, will accept and carry out its obligations under the Act, not only with regard to Gallery Place but with regard to the remainder of the stations in its system. To now set a precedent to the contrary would in this Court's view lead to repeated excuses by defendant that elevator construction has been delayed for any number of facially valid reasons, e.g., lack of funds, construction delays, etc., with a concomitant request to operate the station in violation of the law.[13]

As the court's decision demonstrates, it required strict enforcement to ensure the implementation of technological accommodations that would allow for people with disabilities to fully participate in society. However, a loophole in the legislation itself countered this rhetoric of enforcement. The ABA contained a clause allowing for its standards to be modified or waived on a case-by-case basis when such allowance was "clearly necessary," and the Transit Authority appealed and was granted a waiver under this clause. Although the Transit Authority made the station accessible only two years later, it became clear that the power and enforceability of the ABA was not absolute if economic concerns were sometimes allowed to trump accessibility.

SECTION 504 OF THE REHABILITATION ACT OF 1973

While the ABA only concerned itself with a specific form of access for people with disabilities, the next major civil rights legislation had a far broader reach. Arguably the most important piece of disability rights legislation prior to the Americans with Disabilities Act of 1990, the Rehabilitation Act of 1973 was signed into law by President Richard Nixon on September 26, 1973. This law intended to continue the vocational rehabilitation program established by the Smith-Fess Act of 1920, which provided federal funds to the states for vocational services for disabled citizens. Lawmakers did not intend the Rehabilitation Act to be groundbreaking civil rights legislation; it was meant to be a routine continuation of a previous law. However, at the very end of the act is Section 504: "No otherwise handicapped individual in the United States, as defined in section 7(6), shall, solely by reason of his handicap, be excluded from participation in, be denied the benefits of, or be subjected to discrimination under any program or activity receiving Federal financial assistance."[14] Section 504 was the first federal legislation that granted people with disabilities the same protections and rights as anyone else. The passage stood alone, with no details of how it should be implemented, who would be responsible for its enforcement, or any estimates of the cost of its implementation. Like the ABA, it faced problems with lack of enforcement.

The story of how Section 504 came about is not one of deliberate planning by legislators. The staff of the congressional committee charged with drafting the Rehabilitation Act of 1973 tacked Section 504 onto the end, deliberately mimicking the language of Title VI of the Civil Rights Act of 1964.[15] Staff members felt that without a provision to prohibit discrimination, people with disabilities coming out of the vocational programs sponsored by the Rehabilitation Act would have trouble finding employment.[16] These staff members viewed people with disabilities as deserving the same rights as others to participate in society, and they possessed the power to create actual change. The authors' work went unchallenged; once the Rehabilitation Act reached Congress, there was no debate about Section 504 nor any trouble to get it passed. At the time, Congress did not note its potential significance.

Although the law passed without the efforts of a disability rights movement, public disability activism became involved in the history of Section 504 when the regulations were drawn up by the Department of Health, Education, and Welfare's (HEW) Office of Civil Rights (OCR). The OCR was committed to social change from within the government,[17] and it had experience developing regulations for civil rights legislation, having worked on Title VI of the Civil Rights Act and Title IX of the Education Amendments of 1972. Even so, it took until 1977 for the final regulations for Section 504 to be published. When the OCR began drafting regulations, they called on people with disabilities to provide their expertise, but these meetings were informal and about information, not activism.[18] The OCR contacted individuals with disabilities using personal connections rather than through formal, disability advocacy organizations. Disability organizations were not aware of Section 504 or the OCR's regulations until the spring of 1975, as the initial draft of regulations was being finished.[19] Until this point, the people working to develop Section 504 and its regulations were speaking for the rights of people with disabilities but were not themselves disabled.

Disability activists became more prominent in fighting for Section 504 as the OCR failed to actually publish their final regulations. The American Coalition of Citizens with Disabilities (ACCD), formed in 1974 and led by prominent deaf activist and writer Frank Bowe, held demonstrations throughout 1976, demanding that the regulations be published without being watered down. The OCR waffled and published their draft that spring as a Notice of Intent to Publish Proposed Rules instead of as a finalized document.[20] In response, demonstrations organized by the ACCD continued into 1977 at HEW offices across the country. A sit-in in San Francisco, led by Judy Heumann of Berkeley's Center for Independent Living, lasted twenty-five days and received national attention. Joseph Shapiro attributes part of the success to the reputation of the Bay Area as a center of activism. The sit-in protesters—more than 120 people at the peak—received support from other activist groups, including a gay rights group, the Butterfly Brigade, and the Black Panthers, as well as local government officials.[21] The protesters achieved their desired result, and on April 28 the Section 504 regulations were finally signed.

In the finalized regulations for Section 504, published in the *Federal Register* on May 4, 1977, the Office of Civil Rights attested to the law's significance in the history of disability rights legislation: "Section 504 thus represents the first Federal civil rights law protecting the rights of handicapped persons and reflects a national commitment to end discrimination on the basis of handicap."[22] The regulations make direct reference to the fact that Section 504 mimics the language of Title VI and Title IX. Making more specific the broad language of Section 504, the regulations are divided into seven parts: General Provisions; Employment Practices; Program Accessibility; Preschool, Elementary, and Secondary Education; Postsecondary Education; Health, Welfare, and Social Services; and Procedures.[23] The basic requirements of the regulations are that employers must make "reasonable accommodations" for applicants and employees, unless doing so would cause "undue hardship." Providers of social services must make existing and new facilities accessible to people with disabilities, and any programs must be nondiscriminatory in their selection processes.

As with the ABA, the Section 504 regulations built in an exception concerning economic costs to employers; if accommodations were determined to be unaffordable, then employers could take into account an employee's disability in deciding to retain them. Otherwise, however, the regulations intended to provide access for people with disabilities in many parts of public life. The introduction to the regulations gives justification for creating accommodations on the grounds that a lack of discrimination alone is inadequate to enacting equal rights.

> But eliminating such gross exclusions and denials of equal treatment is not sufficient to assure genuine equal opportunity. In drafting a regulation to prohibit exclusion and discrimination, it became clear that different or special treatment of handicapped persons, because of their handicaps, may be necessary in a number of contexts in order to ensure equal opportunity. Thus, for example, it is meaningless to 'admit' a handicapped person in a wheelchair to a program if the program is offered only on the third floor of a walk-up building. Nor is one providing equal educational opportunity to a deaf child by admitting him or her to a classroom but providing no means for the child to understand the teacher or receive instruction.[24]

Merely prohibiting discrimination could not create a civil rights–based model for people with disabilities. Buildings, workplaces, and educational institutions would have to be physically altered and specialized tools provided in order for people with disabilities to fully participate in society. The regulations acknowledged the costs of such accommodations for recipients, but they provided no outright exemption from the enforcement of Section 504.

The Section 504 regulations were not the only important disability legislation passed during this time. The Rehabilitation Act of 1974 significantly changed the federal definition of "handicapped," which would be later reflected in the heart of the Section 504 regulations. The original 1973 Rehabilitation Act defined a "handicapped individual" as someone who "(A) has a physical or mental disability which for such individual constitutes or results in a substantial handicap to employment and (B) can reasonably be expected to benefit in terms of employability from vocational Rehabilitation services provided pursuant to titles I and III of this Act."[25] The 1974 Rehabilitation Act significantly broadened the definition to anyone who "(i) has a physical or mental impairment which substantially limits one or more major life activities, (ii) has a record of such an impairment, or (iii) is regarded as having such an impairment." The law defined "major life activities" as "functions such as caring for one's self, performing manual tasks, walking, seeing, hearing, speaking, breathing, learning, and working."[26] These definitions expanded the previous focus on difficulties in employment as the defining characteristic of disability to now include other essential aspects of living. The view of society aiding people with disabilities by helping them relieve through employment the burden they imposed had entirely changed. The issue was now about civil rights and equal participation in society.

Education for All Handicapped Children Act of 1975

Furthering the move toward legal civil rights for people with disabilities, the next major civil rights law specifically tackled issues of equal access to education. Requiring that states grant children with disabilities the same rights as nondisabled children, the Education for All Handicapped Children Act of 1975 (EAHC) mandated the inclusion of children with disabilities in public education. This law not only dramatically changed the structure of the public school system but also created a new generation of highly educated people with disabilities who grew up with expectations of fuller participation in society. The EAHC provided for federal funding to the states for education, so long as certain rules were followed—namely, that handicapped children be granted "free appropriate public education."[27] The law required, for the first time, that children with disabilities receive the same level of education as other children.[28] The EAHC had much in common with Section 504 and repeated some of the same requirements regarding K-12 education, although the EAHC was more specific in its instruction.

The effect of the EAHC would become obvious more than a decade later, as the first students who went through the public school system under its requirements began to graduate from high school. As young adults with disabilities graduated into a world where they found fewer protections and rights than they had had in school, a new disability rights movement began to form under a stronger understanding of disability as a social identity and community of people. The beneficiaries of legislation made in the 1970s would go on to fight for the Americans with Disabilities Act of 1990.[29]

Challenges to the Enforcement of Disability Rights

Until the disability rights movement resurged to attain new strength in the late 1980s, it faced a period of challenges and losses. Even when the movement was at its strongest in the mid-1970s, the courts never completely assured the rights of people with disabilities. *Lloyd v. Illinois Regional Transportation Authority*, a lawsuit brought by people with mobility impairments against public transportation services in the Chicago area, exemplified this situation. In this lawsuit, the courts showed continued reluctance to treat discrimination against disabled people as an enforceable issue: "As late as 1976, a federal district judge held that one of the major federal statutes prohibiting discrimination against handicapped persons was merely 'precatory,' a legal term to describe language which entreats, requests, and recommends, as distinct from language which directs and commands. The essence of this judicial holding was that the long-awaited civil rights provision had no teeth in it."[30] Though another court overturned this decision, it is clear that the enforcement of disability rights law stood on shaky ground.

Challenges to the enforcement of Section 504 continued with another landmark case in 1976, *Davis v. Southeastern Community College*. Frances Davis, a licensed practical nurse for almost ten years, sued a local college for refusing her admittance into their registered nurse program on the basis of her hearing impairment. The court ruled against her, claiming she could not satisfy the duties of an RN, notably in emergency situations.[31] Davis appealed the ruling, and the Fourth Circuit Court of Appeals sided with her, concluding that the college could not take her disability into account when deciding admission. The college challenged this ruling before the Supreme Court, who overturned the Court of Appeals. They judged that the nursing program could consider physical qualifications alongside academic and technical abilities and that the college was not obligated to adjust their curriculum to create a track that would work for Davis.[32] Activists criticized the ruling for failing to defend Davis and the court for not defining the limits of its ruling and how it should apply in similar cases, thereby leaving enforcement of Section 504 open to other challenges. Section 504 could not, on its own, guarantee people with disabilities equal rights and protections against discrimination.

In many ways, the influence of the disability rights movement faded during the decade following 1978. Attempts by activists to extend Title VII of the Civil Rights Act to include discrimination on the basis of disability in all types of employment failed to find congressional supporters. There were few legislative advances for disability rights after the publication of the Section 504 regulations.[33] At the time Kent Hull's book went to press in 1979, he feared that there was inadequate enforcement of existing laws, as well as threats to ground already won for disability rights.[34] Part of the reason for this lack of progress was geographical; other areas of the country lacked the activist culture of the Bay Area and were reluctant to take on the costs associated with accessibility. Unlike the extension of civil rights to nondisabled groups, antidiscrimination for people with disabilities cost money.[35] The disability rights movement would not regain its strength until the late 1980s and the move to pass the Americans with Disabilities Act.

Computer Technology and People with Disabilities

As people with different kinds of disabilities found employment using computers with technological accommodations or were the human subjects of cutting-edge research, technology users and developers realized the potential computers held to change what it meant to be disabled. Disability access and computer technology became experimental sites for each other as new ideas of usability and imagined users were conceived. Here I examine accessible technology and people with disabilities in four aspects of the computer industry: blind professionals, a national organization for professionals, computer science research, and a large-scale technology company. In addition to developing assistive devices that allowed computer professionals with disabilities to perform their jobs, technologists and researchers also anticipated the potential of the computer to one day be a technology that could change people's lives for the better, and in particular, could benefit people with disabilities.

BLIND COMPUTER PROGRAMMERS IN THE 1960S AND 1970S

There existed a brief period during the 1960s and 1970s when the state of computer technology and views on the capabilities of people with disabilities created a situation that encouraged blind people to become computer programmers. During the late 1960s, the Association for Computing Machinery (ACM)—the largest professional organization in the world for people who work on computers—ran a Committee on Professional Activities of the Blind. The committee started in 1964, published a newsletter for four years, and organized a conference in 1969. In an article published by the ACM in 1964, the chair of the newly formed committee, Theodor D. Sterling, and his coauthors described the possibilities for blind people working as programmers.[36] Addressing employers more than potential employees, the authors argued that blind people not only were capable of being programmers but were particularly well suited to the job; this news could benefit employers, as there was a shortage of programmers at this time. The authors argued that blind people might be inherently skilled at programming because navigating an environment nonvisually requires an understanding of space and organization that is applicable to understanding the layout and operation of large, complex programs.[37]

In addition to these abilities of "orientation," blind people required few, inexpensive technological accommodations to work in computer programming. First, for program preparation, blind programmers were unlikely to need to punch their own cards for the program, as this was a job usually done by clerks, not programmers. Computer work was organized during the 1960s with a programmer designing programs and writing out instructions, which a clerk then translated onto punch cards that were read by a mainframe computer. The clerk therefore performed the actual operation of the computer. In order to provide the clerk with the program, a blind person could either type in and print out the program (a slower method) or, using special paper with a diagram of a punch card embossed onto it, the programmer could feel where instructions should be placed on the punch card and write them in (a more expensive method). The authors argued in favor of the latter method in spite of the additional cost of specialized paper.[38]

Second, for program assembly, the programmer needed to read printouts of program listings and memory dumps. These could be translated into braille and printed using a standard printer, with the only additional cost being the braille translation software.[39] Third, concerning program execution and debugging, the authors discussed two options. For simple programs, the programmer could print memory dumps in braille and read them to find errors. For more complex programs, the programmer would need to read the position of various knobs and buttons on the computer console (which a blind person could do by feel) and see the activation of lights. The authors suggested that a blind programmer could use an inexpensive "heat sensitive probe which translates the light into sound" in order to hear the console lights.[40] Finally, a blind programmer would occasionally need to read a punch card itself, which could be done via a special mechanical reader that allowed the user to feel and interpret the card.[41] All of these technologies were listed as relatively inexpensive and easy to use.

This accommodation of the needs of blind programmers anticipates the coming civil rights model of disability by showing that participation is possible through technological accommodation and benefits both employees and employers. In other ways, however, the treatment of these early blind programmers demonstrated their lack of full equal rights because it regarded disability as a problem that individual people have and that must be fixed enough to allow them to participate but that otherwise must not get in the way of their jobs. In 1966, the ACM's Committee on Professional Activities of the Blind published *The Selection, Training, and Placement of Blind Computer Programmers*.[42] In addition to repeating technical information similar to that found in Sterling's article two years previously, this short manual also demonstrates the discriminatory and condescending work environment blind people faced in the late 1960s. Prior to any legislation providing civil rights for people with disabilities, the responsibility to fit into a sighted workplace fell solely on the blind person: "The blind candidate for employment has to satisfy the manager of the center that his handicap will not interfere in the smooth performance of his own work or that of his co-workers."[43] Blind employees had to ensure that they did not get in the way of their sighted colleagues with any accommodations they might need to perform their jobs or move about the workplace: "It is the responsibility of each person to make sure that the use of these special devices and aids or dogs does not impinge in any way on the rights or interfere with the convenience or safety of his co-workers."[44]

This ACM publication intended to encourage blind computer programmers to join the profession, but the book's tone makes it clear that, in some ways, society was not ready for people with disabilities to publicly take part in everyday activities: "He should have successfully resolved any problems related to his blindness which might stand in the way of his training or employment. Unpleasant mannerisms sometimes associated with blindness must be corrected if he is to secure and hold gainful employment."[45] If blind programmers could accommodate themselves to the established operation of sighted workplaces—and not appear overtly disabled—then they could find a place in a technical field where the kind of work and tools existed that required few alterations for them.

While punch-card computing remained common through the early 1970s, the technology that blind programmers used changed little. In 1973, a newsletter run by the successor to the ACM Committee on Professional Activities of the Blind printed the results of a survey given to blind programmers.[46] The results do not indicate how many people responded (although, to give a rough idea, open-ended questions had between forty and forty-seven answers), but they do describe what assistive technologies blind programmers used, including braille writers, tape recorders, typewriters, punch-card readers, braille software, and other related tools. A second question asked respondents what technology they wished existed; answers mostly involved improvements to current technology, such as faster conversion of print to braille, faster punch-card readers, and ways to combine various technologies so that programmers would not have to use multiple tools. Ninety percent of respondents stated that a faster punch-card reader would be valuable to them, indicating a shift from the 1960s when the ACM described programming as a job that required little time spent reading actual punch cards. A more general question asked what other developments blind programmers would like to see in their occupations. Many responses mentioned improved education and training for blind people, as well as the education of employers and sighted programmers about blind programmers' abilities in order to make gaining employment easier. Respondents also desired greater availability of computer manuals in braille.

The trend of employing blind people as programmers dropped off in the 1970s due to changes in computer technologies and workflow operation. As punch-card readers gave way to keyboard terminals, blind people no longer found easy adaptive devices to create and read programs. Increasingly, programmers worked on computers themselves, without going through the intermediary of a clerk. Instead of reading output through a printout or lights on a display, programmers needed to control a computer through the use of a monitor screen. Different kinds of assistive technologies, such as screen readers, would be needed for blind people to operate these new machines. The shift from punch cards to terminal computing exemplified the kind of shift in standardization which can leave people with certain types of disabilities behind. The previous technology worked with the needs of blind people, via simple accommodations, but they were also required to hide their disability as much as possible. Now the new computer paradigm excluded blind people altogether, operating counter to their needs. The drive toward integration was proving slow.

Special Interest Group on Computers and the Physically Handicapped

As the concept of integration became more normalized, the attention toward computer technology for people with disabilities spread. For the ACM Committee for Professional Activities of the Blind, the focus broadened to include other people with disabilities, and the committee transformed itself into SIGCAPH (Special Interest Group on Computers and the Physically Handicapped). Special Interest Groups (SIGs) are a way for ACM members with common interests to connect and communicate with each other; the groups publish newsletters for their members and often run their own conferences and workshops. ACM members created SIGCAPH in 1971 as a way for computer professionals with disabilities and people who worked with and supported them to communicate with each other about current research and technological developments. It was divided into sections for the blind, deaf, and motor impaired. For two decades, it was often the smallest of all the ACM SIGs, sometimes with fewer than three hundred members.

Following its inception, SIGCAPH experienced periods of enthusiasm and productivity, but it also struggled for twenty years to convince members to volunteer as officers or contribute materials to the newsletter. Individuals kept the group alive during its downtimes; various editors often contributed much of the newsletter materials, and some chairpersons stayed in office for the maximum time allowed by the ACM. Even when early personal computer development started to take off, SIGCAPH was not in a position to cover the new advances. It is unclear why SIGCAPH members were mostly apathetic about the state of the group for so long, although lack of a sense of community seemed to play some role.

SIGCAPH was not formalized as an activist group but as a professional organization, and, as the computer was some years away from being a consumer technology, the members were not public users. Yet the group provided similar functions for its members as the disability and computer technology activist groups that later emerged in the 1980s. Aside from offering a way for members to communicate with each other and learn about technological developments, SIGCAPH also worked to promote the education and hiring of computer professionals with disabilities. The founding goals of SIGCAPH, in 1971, were:

1. To promote the professional interests of computing personnel with physical disabilities.
2. To promote the application of computing and information technology toward solutions of disability problems.
3. To perform a public education function in support of computing careers for suitably trained blind, deaf or motor impaired persons.[47]

SIGCAPH added a fourth goal two years later, in 1973.

4. Promote the interest of professionals by:
   a. Affording opportunity for discussion of problems of common interest.
   b. Encouraging presentation of papers of special interest to this group at annual and Regional Meetings of the ACM and at other special meetings organized by this group.
   c. Providing guidance to the ACM Council on matters of importance to this group.
   d. Publishing a newsletter containing information of interest to this group.
   e. Other appropriate means.[48]

The goals of SIGCAPH mostly focused on the promotion of computer professionals with disabilities. Especially in the early years, the newsletter reflected this mission by printing information on educational programs geared toward people with disabilities, technical articles on specialist technologies used by disabled professionals, and letters from or accounts of the experiences of people with disabilities working in computer-related occupations. In addition to professional concerns, however, the group also promoted research and development of specialized adaptive devices for people with disabilities, as well as general accessible technologies that allowed more people to use computers.

Around the same time the Office of Civil Rights was drafting the regulations of Section 504, another office in the Department of Health, Education and Welfare, the Social Security Administration (SSA), asked SIGCAPH members to contribute their knowledge and experience to help develop an accessible government-run computer center. In late 1973, the SSA asked SIGCAPH members to review the new center's building specifications, so that architectural barriers would not prevent access.[49] In July 1974, the SSA reported on the successful creation of the computer center. In a letter to SIGCAPH, the president of Computation Systems—the company hired to write the building specifications—commented on its member involvement: "To exemplify the kinds of value that can be added to a specification by people who really know and care, let me mention just one item: In the comments that came from one SIGCAPH member (on behalf of the Cleveland Institute for the Blind) there was marked, in the section on walks and ramps, the three-word marginal note, 'no side slopes.' Nowhere in the documents we have seen does there appear this simple but important specification."[50] In order to build in accessibility, designers must be aware of the needs of people with different kinds of disabilities and how they use assistive technologies. In this case, building gradual slopes on the sides of walkways would have made it difficult for blind people to tell which level surface they were walking on. By working with people who had expertise, this computer center was able to include architectural details that were not yet part of standardized guidelines for accessible design.

However, as with organizations required by federal law to become accessible, SIGCAPH's own commitment to accessibility was sometimes hampered by economic concerns. The fact that accessible technologies are differently usable for users with different disabilities complicates efforts at universal design for computer technology. For example, SIGCAPH sought a balance between economics and accessibility in their choice of what version of their newsletter to supply to visually impaired subscribers. When the group formed, they decided that the braille version should not cost members more than the text version, on the basis "that there should be no distinction in membership fees because of handicap. Any slight difference in the cost of print or braille editions should be borne by the full SIG membership, just as well as it should bear the cost of interpreters for the deaf at business meetings, special meetings, or conferences as agreed upon by the SIG."[51] The group maintained the braille edition of the newsletter for eight years, until the costs finally became infeasible. SIGCAPH first considered dropping the braille version and replacing it with audio tape in 1978, as braille then cost five times more than audio tape.[52] Norwegian member Kari Larsen argued against this change: "The information explosion we experience every day, makes it necessary to read exactly those articles that are of major interest to us. We know that blind people is [sic] very far from obtaining all the written information that seeing people get. . . . As an experiment, why not send printed versions of the newsletter only to deaf persons, and cassettes to everybody able to hear? May be some of us will find that it takes too long time [sic] to listen to the newsletter and that it is too difficult to find the interesting articles?"[53]

Choosing between these two ways of delivering the newsletter was a matter not only of economic concern but of usability; the practice of reading differs with these different technologies. As Larsen explains, for a blind person, it is far more difficult to skip around an audio tape than a braille document, and it is more time consuming to listen rather than read by touch. SIGCAPH's budget supported the braille edition for a couple more years before the group finally discontinued it in 1980. By then it cost almost ten dollars per copy. Rather than charging members who wanted a braille version more than those receiving the text newsletter, the group replaced the braille version with an audio-tape version.[54] It opted to change the form of the accessible technology they used in order to accommodate their visually impaired members equally in terms of cost, even though the new technology was less usable for some. Situations requiring a balance to be struck between economic concerns and differences in usability for different people arose repeatedly throughout the development of accessible personal computer technologies.

Murray Turoff and Computerized Conferencing

One of the first computer technologies specifically created to benefit people with disabilities while also being designed for more general use was computerized conferencing. In 1975, researcher Murray Turoff gave a talk at a meeting of the American Association for the Advancement of Science on early work on this technology, a predecessor to later online communication systems, such as message boards and chat rooms. Computerized conferencing allowed multiple people to communicate together online. Participants could write to each other at the time of their choosing, with all messages in the group conversation stored online on a central server. Turoff describes the unique properties of this technology as a communication medium.

1. The individuals no longer have to be coincident in time, as in telephone calls or face-to-face meeting, since the computer keeps a record of the discussion and a bookmark for every individual on what he has seen.

2. The system allows each individual to work at his own pace, taking as much or as little time as he wishes to read, contemplate and/or reply (i.e., a "self activating" form of communication).

3. The system provides many of the signals present in face-to-face communication, i.e., who is in the discussion at any particular instant, what everyone has seen or not seen, when they were last in the meeting, etc.

4. The system provides a host of unique features, i.e., private messages or whispering between individuals, items that can be voted on, specialized retrieval—key words, authors—to reorder the discussion, conditional messages, etc.[55]

Computerized conferencing allowed for communication between people without some of the limitations of face-to-face conversation, as participants could interact with each other on their own time and in their own space.

Turoff believed this aspect of the technology would make it particularly useful for people with disabilities: "It is the view of the author that this type of communication offers tremendous potential for improving the opportunity for [people with disabilities] to lead more rewarding lives and to decrease greatly the limitations often imposed upon their mental capacity by the presence of inhibiting physical disabilities."[56] The disabled user is the imagined or ideal user here. This technological system designed with the needs of people with disabilities specifically in mind allows it to be usable by everyone, while providing accommodations for people with different abilities to participate in a new form of social interaction. The following year, Turoff began the trials he hoped for, asking for interested groups of computer users with disabilities to test a new system.[57] While computerized conferencing as a specific technology did not endure as Turoff and his fellow researchers imagined, it was a predecessor to now ubiquitous networked communication technologies. The prominent place of people with disabilities in the writings on computerized conferencing makes this an ideal case for examining the computer as a universalizing technology of augmentation.

The most prominent publication of Turoff's research was his 1978 book *The Network Nation: Human Communication via Computer,* which he cowrote with sociologist and computer researcher Starr Roxanne Hiltz, explaining what they saw as the future for computerized conferencing and its potential effect on society.[58] In the preface, the authors write, "We believe that [computerized conferencing] will eventually be as omnipresent as the telephone and as revolutionary, in terms of facilitating the growth and emergence of vast networks of geographically dispersed persons who are nevertheless able to work and communicate with one another at no greater cost than if they were located a few blocks from one another."[59] This technology would offer more than a new means of communication; in their view, it would organize new networks of human interaction across the world. These networks would be based not on geographical proximity but on connections unbound by physical location: "We will become the Network Nation, exchanging vast amounts of both information and social-emotional communications with colleagues, friends, and 'strangers' who share similar interests, who are spread out all over the nation. . . . we will become a 'global village' whose boundaries are demarcated only by the political decisions of those governments that choose not to become part of an international computer network. An individual will, literally, be able to work, shop, or be educated by or with persons anywhere in the nation or in the world."[60] Hiltz and Turoff's predictions for the future of computer technology turned out to be highly accurate. Built into this view of the future are the implied values of the computer as a universalizing technology: a technology for any purpose that can unite users across the world and break down geographical boundaries. At the same time, the computer becomes the necessary means by which to accomplish such a "global village," as it permits people to overcome the physical limitations that prevent such networks from existing without computer mediation.

People with disabilities, in particular, were a group Turoff and Hiltz saw as benefiting from computerized conferencing technology. The authors write that, "the biggest advantage of computer-mediated communication is that it spans space and time barriers, allowing a person to work, learn, and communicate from those places and at those times that are most convenient for him or her. Thus the mobility limitations of the physically handicapped make them a disadvantaged group that can benefit greatly from this technology."[61] Turoff and Hiltz envisioned computer-mediated communication as a technology that was designed to take into account use by people with disabilities, similar to a building designed to be barrier-free so that anyone can access it. Computerized conferencing created a space for social interaction that included people with disabilities as intended users. This new form of communication augmented all users' abilities to interact in a new social arrangement; it created networks independent of space, as well as a means of communication that was not instantaneous and therefore less dependent on time.

Turoff and Hiltz continued their research on computerized conferencing at the New Jersey Institute of Technology (NJIT). In 1979, the SIGCAPH newsletter reported on their research, specifically on the use of computerized conferencing to connect elderly women with children who had cerebral palsy. As part of a grant funded by the National Science Foundation (NSF), the program brought women at a nursing home together with children at a Cerebral Palsy Center to communicate with each other via a computerized conferencing system. Led by Turoff, the focus of this study was on the emotional well-being of the participants using the technology: "The NJIT scientists are convinced that the continual availability of someone eager to 'listen' can bring a new meaning to life, both for the handicapped children and for the elderly women."[62] The researchers chose these two groups of people because of their relative isolation and dependence on others for caretaking, with the idea that these women and children might desire communication with people outside their restricted environments. Computerized conferencing was particularly useful here because the speed at which the participants typed did not matter to the system.[63] For these researchers, computerized conferencing could help to create a more level playing field, where one's disability did not affect one's communication with others. Users with disabilities were the test case, the initial users whose needs had to be met before the technology was then generalized for all users.

Hiltz and Turoff explicitly viewed computerized conferencing as a technology of human augmentation: "a goal of [computerized conferencing] systems is augmentation of communication processes by the presence of the computer."[64] For people with a disability that affected face-to-face or telephone-based communication, computerized conferencing promised a means of communication where disabilities were potentially unseen. The authors describe this feature explicitly: "various participants need not be aware of the disability that any of them suffer unless a person wishes to volunteer the information. . . . Even if the participants are aware that a particular person is blind or deaf, the social salience of the characteristic is much less, because it is not visible. Moreover, if it takes a handicapped person longer to read and/or write into a system, this does not slow down or inhibit the speed or ease of participation of the other members."[65] The authors promoted computerized conferencing as a way not only to grant someone abilities they did not otherwise possess but also to cover the disabilities they had. Compared with the case of the blind computer programmers, who were required to prevent their disability from negatively affecting their coworkers, the users in the computerized conferencing experiment were offered a nonprescriptive option to choose whether they revealed their disabilities, along with any other aspects of their identities, to others. There was liberatory potential in not having to hide or show themselves but also the possibility of invisibility and assumptions of normativity by other users.

One of the goals of computerized conferencing was to remove a person's disability from the act of communication, in so far as possible, both by making participants unaware of each other's disability status and by providing a communication system relatively unaffected by disability. This research explicitly imagined people with disabilities as users of the technology; making it work for people with different physical needs was a part of the propulsion for innovation in creating a cutting-edge technology and new form of communication. Computerized conferencing was about more than just accommodation; it challenged ideas of normalcy and showed that personal technologies and new social environments could be developed with the needs of different people in mind.

IBM AND PEOPLE WITH DISABILITIES

When IBM was founded in 1911, it was called the Computing-Tabulating-Recording Company and sold mechanical tabulating and accounting machines. In 1924, the company's president, Thomas J. Watson, renamed it International Business Machines Corporation,[66] and with the invention of the digital computer in the 1940s, IBM turned its attention toward mainframe computing. IBM dominated the computer industry during much of the mid-twentieth century, controlling around 70 percent of the market from the late 1950s through the 1970s.[67] As it rose in power, however, the company hung onto some of the progressive values that Watson and his son had instilled. These values included hiring practices that promoted diversity across gender, race, and disability.[68] That IBM hired its first employee with a disability in 1914 was a continuing point of pride.[69] In 1944, the company's efforts to hire people with disabilities were recognized by a congressional subcommittee as, according to IBM, "a shining example other companies might follow."[70] During World War II, its efforts to hire and train people with disabilities were motivated by the need to grow a workforce that had been diminished by soldiers leaving, and after the war the company needed to make accommodations for disabled veterans.[71]

By 1977, IBM was able to feature a number of employees with disabilities in its company magazine, *Think*. This sampling included people with various disabilities who worked in management, programming, administration, and engineering.[72] Of the seven employees featured, three worked on projects to develop assistive technologies or training for other people with disabilities. A decade later, in 1988, IBM employed around 7,000 people with disabilities among their global workforce of more than 387,000.[73] That same year, the federal President's Committee on the Handicapped named IBM as Employer of the Year for its long history of employing people with disabilities.[74] The company's long-standing dedication to employees with disabilities appears to be a part of its larger practice of supporting a progressive workplace through diverse hiring standards and enforced antidiscrimination policies.

IBM made numerous alterations to make the workplace accessible for all employees. A *Think* article from 1988 described some of these accommodations, including removing architectural barriers and making changes to buildings, providing captioning for videos and sign language interpreters at meetings, offering audio versions of bulletins and *Think* magazine, and making workplace technologies accessible through the use of adaptive devices.[75] Notably, these workplace accommodations were made two years before the passage of the Americans with Disabilities Act required such changes for employees with disabilities. Before the legislation was passed, IBM was already hiring employees with disabilities and adjusting the workplace so that they could perform their jobs the same as their nondisabled colleagues.

IBM also ran programs to train nonemployees for computer careers. In 1968, the company created a rehabilitation department in its Office Products Division. Bert Williams, the product manager in 1972, described the department's work at a time when they found no similar programs on which to base theirs: "Our goal is not just rehabilitation, but to help each individual realize his or her maximum potential. There were no precedents for us to follow in this work. . . . Our people act as catalysts, calling on organizations where we feel we can help. They help define a need and develop a program to meet it, using the training resources we've already developed, and then support the local community in making it work."[76] This perspective on disability in IBM's rehabilitation department anticipated activists' arguments in the 1980s that it was possible for personal computer technology to enable people with disabilities to pursue their goals. Williams also described IBM's emphasis on training people to use computers rather than just donating technology to charities and individuals: "We receive many requests for machines to be used in training handicapped people. We feel that the real need is not for donated machines, but rather business involvement with good programs that have sufficient skills development to insure pursuit of a career path in a job that is guaranteed at the end of the course. Now we know that it's the total program that counts—not just handing out machines."[77] IBM helped people with disabilities find work in computer-related careers through training programs the company organized; these resulted in thousands of people placed in jobs that allowed them to utilize the kinds of technical skills that Williams discussed.

In 1972, IBM started its Computer Programmer Training for Severely Physically Disabled Persons program, based on a suggestion from an employee with disabilities.[78] By 1988, there were thirty centers across the United States where people with disabilities could receive training in computer programming. During this time, 2,400 people graduated from the training program, with an 80 to 85 percent success rate in finding jobs.[79] The program later broadened its focus in terms of training and who could be admitted into the Personal Computer Based Skills Training for Disabled Persons program, leading to three thousand graduates from forty centers by 1996. Beyond computer programming, the program added software training in "word processing, data entry, desktop publishing, and computer-aided design."[80] With the introduction of the personal computer, IBM's emphasis on employment for people with disabilities changed to a focus on personal use and empowerment.

In analyzing the history of disability rights legislation and the use of computers by people with disabilities prior to the personal computer, I introduced five historical themes to which I return throughout this book. First, disability rights underwent a shift during the late 1960s from a paternalistic, caretaker model to a civil rights–based model. The view of people with disabilities slowly changed from their being regarded as a burden on society that they could try to alleviate—by taking part in vocational programs in order to find gainful employment—to their being seen as having personhood and the fundamental right to full participation in society. The struggle for civil rights for people with disabilities coincided with the rise of identity politics in the United States, and disability became a social identity and the basis for a movement comprised of a population fighting for equality. Federal legislation, such as the Architectural Barriers Act of 1968, Section 504 of the Rehabilitation Act of 1973, and the Education of All Handicapped Children Act of 1975, guaranteed the civil rights of people with disabilities. With this change, people with disabilities were less often segregated into special programs, isolated from the rest of society, and instead were slowly integrated into mainstream programs with everyone else. Throughout this shift, the rights of people with disabilities have historically been spoken for by different groups—activists, lawmakers, judges, employers, and, of course, people with disabilities themselves.

Second, in order to fully participate in society and have equal rights, people with disabilities required technological accommodations. Public space and many technologies are designed for use by people without disabilities; therefore that space or technology has to be altered to allow full access. Technological accommodations included removing architectural barriers and creating assistive technologies that allow participation in school or work. The economic cost of technological accommodations affected their enactment historically, even when federal law required institutions to provide those accommodations and institutions desired to enact them.

Third, the personal computer as a technology of accommodation was influenced by the values embedded into it, in particular, its function as a universalizing technology—a tool for anyone, for any purpose. In spite of these values, computer users with disabilities required specialized technologies in order to reach the potential the computer promised. For blind programmers in the 1960s, working on mainframe computers, these accessible technologies took the form of small devices or changes to the work process that allowed them to do their jobs. For personal computers in the 1980s to be usable by all, the development of accessibility was a complex interaction between computer companies and users with disabilities.

Fourth, in addition to promoting its value as a universalizing technology, the early personal computer carried with it the idea that it was a technology of augmentation, that it allowed people to expand their intellectual abilities, which were limited by their bodies. The computer provided abilities that people do not possess on their own. Computerized conferencing and later networking created new forms of communication that allowed people to interact in their own time and space. From the perspective of the computer as an augmentation device, everyone is disabled; it is inherently an assistive technology.

Fifth, understanding the computer only as a technology of augmentation that makes up for limitations all people possess misses the lived reality of its use. The computer can never provide a completely level playing field where everyone has the same opportunities and abilities because it is still a machine used by people with bodies. Differences in those bodies—such as the presence of a disability—matter to the person using the technology. Even if the guidelines of universal design are followed, the computer can still never be used in the same way by all people. There are differences between the purpose of a technology for its developers, the values imprinted in it, and its use.

In the twentieth century, the United States witnessed a profound change in the lives of people with disabilities. Society went from viewing them as helpless and needing paternalistic charity to seeing them as people in their own right, with goals and the desire to participate in the world as everyone else does. Legislation began to grant them rights, and technological accommodations began to remove some obstacles. The development of computer technology created jobs and the possibility of future accessible technologies. With the birth of the personal computer in the late 1970s, the potential of computer technology was realized with the creation of machines anyone might own at home. First, though, technical knowledge needed to be disseminated to those not on the forefront of innovation. Just as with buildings or public transportation, adaptations would need to be developed to make computers usable for everyone.

Chapter Two

Early Personal Computer Accessibility, 1980–1987

The personal computer began to be available as a consumer product in the late 1970s. These early machines were limited in usability and functionality, yet carried with them embedded values of the computer as a universal tool and a means of human augmentation. The potential to change people's lives for the better spurred immediate tinkering with the technology to improve it for use by people with disabilities, but a lack of standardization and the absence of social technologies to transmit awareness meant that it took until the mid-1980s for accessible computer technologies to begin to be readily available, affordable, and easy to use. In this chapter, I trace the evolution of accessible personal computer technologies, during the early-to-mid-1980s, from their birth in entrepreneurial tinkering, to their development as consumer technologies by small start-up companies as well as large-scale computer corporations. I also examine how personal computer technology was taken up and promoted by people with disabilities and activists.

The personal computer led to a new networked form of production between users, developers, and the technology itself. This network allowed for the creation of new social spaces of interaction, where groups of people with common identities could come together. People with disabilities comprised numerous subgroups of people with very different needs that specific accessible technologies were created to meet. Universality became an increasingly important value and led to attempts to accommodate users by meeting all their individual needs. Technological accommodations, during this time, also demonstrated the effects of applying market solutions to social problems, including such limitations as technology being unable to effectively reach users due to issues of perceived limited market and struggling small businesses.

Computer Potential

In the late 1970s and early 1980s, the promise of computer technology for people with disabilities was still unrealized. The personal computer had the potential to be a universal tool for any purpose as well as a technology of human augmentation. First though, developers needed to set standards that would allow people with different needs to utilize the technology. In addition, social technologies—communication and organizational networks—had to be built in order for the liberatory potential of the personal computer to manifest.

Researchers viewed the early personal computer as a tool that offered access to a nonphysical public space. Even in the early 1980s, people anticipated the possibilities of conducting one's everyday life online though the personal computer. This possibility was especially promising for people with mobility impairments, as the computer could potentially allow them to accomplish public business, such as shopping or money management, without having to leave home. At this time, the promise of the personal computer was that it might give people with disabilities the same opportunities in society that nondisabled people enjoyed. Optimism reigned. In 1984, Peter McWilliams wrote in his book *Personal Computers and the Disabled,* "Personal computers can make the difference between communication and isolation, between productivity and non-productivity, between independence and dependence."[1] In *Microcomputer Resource Book for Special Education,* also from 1984, Dolores Hagen spoke of the role the personal computer had played in her son's life: "Telecommunications via the microcomputer will, for the first time, give the handicapped equal opportunity in society."[2] The promise of the personal computer as a tool to solve people's problems and open up the world originated from the counterculture-hobbyist environment of Silicon Valley.[3] For early developers, the personal computer acted as a successor to the *Whole Earth Catalog,* exemplifying the same goal of being a tool that provided a means for people to create the kinds of lives they chose to live. Whereas the catalog offered all the physical objects that would permit people to build a new society, the personal computer provided the means for people to create their own digital forms of problem solving and social interaction.

However, early personal computer developers did not design their technology with people with disabilities in mind, so it first had to be made accessible. Gregg

Vanderheiden, of the Trace Center at the University of Wisconsin–Madison,[4] drew attention in the early 1980s to this need: "Very rapidly, our society is moving toward electronic assisted everything. In the process, electronic pathways are being laid throughout our society—pathways which could tremendously increase the functional mobility and capabilities of individuals with physical and sensory disabilities. All of these electronic information pathways will be of little use, however, if unrestricted access is not available. Patching one or two access points is not sufficient, in the same manner that providing curb ramps or curb cuts for some of the sidewalks is not sufficient."[5] His main argument was that being able to make one's home computer accessible—and have access to all the possibilities the computer offered from home—was not enough; it would be like only having curb cuts on the sidewalks around your block. Vanderheiden predicted that computers would quickly become necessary in all aspects of public life; everything from schools to workplaces to banks to government offices would soon be run by computers. If people could access their own computers, after modifying them with adaptive devices and specialized software to fit their own needs, but still could not access public machines, then the computer revolution would do nothing but create new barriers. As a personal and a public technology, all personal computers needed to have some means of accessibility built into them to accommodate the needs of different bodies if the machines were going to grant access to a new digital world for everyone.

In the early 1980s, personal computer technology was still fairly undeveloped. While accessibility features were not commonplace, a lack of accessibility had not yet become standard. Until the mid-1980s, however, large computer companies showed little interest in developing their own accessible technologies. Deaf activist and former head of the American Coalition of Citizens with Disabilities Frank Bowe discussed the need for computer manufacturers to make it standard for their technologies to work for people with disabilities: "Just as buildings had to be made accessible before physically disabled and older people could use them, so too will computers have to become accessible before special-needs persons can become full partners in the computer revolution."[6] Unlike buildings designed and constructed only with able-bodied users in mind, computers were still unfixed enough that a lack of accessibility could be undone before standards were set and the technology stabilized.[7] By including accessibility concerns in the development of personal computers early on, the technology would not be made disabling. Vanderheiden likened this to creating curb cuts while the sidewalk is still being laid, instead of waiting until later and having to tear up the concrete to build in accessibility.[8] Developers needed to anticipate and understand different uses of computer technology by people with different bodies, in order for the computer to not be standardized with obstacles built in.

Johns Hopkins's Contest on Personal Computing to Aid the Handicapped

Researchers and innovators tinkering with personal computer technology began to make good on its potential for people with disabilities in the early 1980s. Prior to most standards being set, these technologists created prototypes of machines that would, in only a few years, become consumer technologies enabling people with disabilities to use personal computers. A number of prominent innovators took part in a competition in 1980 and 1981 to reward people seeking to develop computers so as to benefit of people with disabilities: the First National Search/Contest on the Application of Personal Computing to Aid the Handicapped, run by Johns Hopkins University's Applied Physics Laboratory, with sponsorship from the National Science Foundation and Radio Shack. The contest brought together the top regional submissions for a final national contest. The top three national winners were awarded $10,000, $3,000, and $1,500, respectively, while seven honorable mention winners received $500 each.[9] These winning inventors and their technologies offer a broad view of who was working on computer accessibility at the time, the kinds of personal computer technologies available, and their potential application for people with disabilities. This set of winning technologies not only provides a glimpse at early versions of accessible computer technologies that would go on to become common consumer products but also offers examples of the different kinds of embodied uses personal computer developers needed to understand if the computer was going to be usable by everyone.

A notice of the contest in the ACM SIGCAPH newsletter described its two objectives, which were to "focus the power of computing technology on the urgent needs of millions of handicapped citizens" and to "harness individual innovation & creativity on a national basis."[10] The capabilities of computers to help people with disabilities were seen as "virtually unlimited."[11] The contest divided entrants into Professionals, Amateurs, or Full-Time Students and categorized their entries as Computer Based Devices (hardware), Computer Programs (software), or System Concept/Design (ideas with implementation). A key rule of the contest was that off-the-shelf components, with modifications, were required, so as to demonstrate the possibilities for consumer technologies to be adapted with relatively little tinkering for use by people with disabilities. The competition defined disability broadly as "any limitation of functional capabilities including mobility, communication, self-care, and self-direction."[12] Computers, through the creativity of inventors, could help improve anyone's life. The contest's director, Paul L. Hazan, described this potential: "From aids to independent living to flexible tools that can greatly increase the variety and quality of job opportunities, the rapidly evolving field of computing is pregnant with possibilities."[13]

The plethora of technologies intended for use by people with widely different disabilities featured in the contest indicates two salient aspects of personal computer technology and accessibility at this time. First, it offered a wide range of possibilities; personal computers promised to be a technology for any imagined use, and limitations that existed one day would be overcome by advances the next. The contest's requirement to use as near to off-the-shelf components as possible shows how even early personal computer technology could be adapted to uses beyond those explicitly intended. Second, accessibility and the use of personal computers by people with disabilities were, for the most part, not an integral part of the design of early personal computers. Devices had to be adapted, components altered, and software written in order for people with varying disabilities to use this technology.

The first-place winner of the contest was Harry Levitt from City University of New York. He used a TRS-80 Pocket Computer and a modem to create a mobile communication system for deaf people which would operate over public phone lines—a low-cost, improved version of the teletypewriter (TTY) already in use. Levitt's Portable Telecommunicator for the Deaf allowed users to transmit messages via a computer keyboard, store the messages in memory, print messages, and read from an audio cassette. He saw a need for his device in the way hearing and deaf phone users interacted differently with the phone system. Hearing users were able to make urgent phone calls in public places using pay phones, but deaf users had no real access to pay phones and lacked a portable communication system. Levitt's Telecommunicator could replace TTY systems, which were heavy, expensive, slow, and had no means to store messages.[14] Levitt also saw his device as a stepping stone to future technologies that would benefit deaf people: "Perhaps the most important advantage of all is that the use of a pocket computer as a convenient, inexpensive communication device introduces the deaf telecommunicator user to the concept of an intelligent, computer-based communications system of almost unlimited scope and flexibility."[15] He hoped that by using commonplace personal computer technology barriers between deaf and hearing people could be reduced.

The second-place winner, Mark Friedman of Carnegie-Mellon University, developed a communication system for users who had mobility impairments and were unable to speak. He and his co-researchers created the EyeTracker for Communication, which used an infrared camera to follow eye movement and detect where a user was looking on a computer screen. The screen displayed words in eight regions on the screen that the user could either select as final output or use to bring up another eight related, more specific words. The computer read out loud the words the user selected, using a voice that was gender and age appropriate. The system was intended to be used by children who did not have steady control of another body part, with the understanding that eye movement would be less fatiguing for the user and faster than trying to control a mechanical switch.[16] The researchers initially ran the EyeTracker on an Apple II computer for use as a teaching aid and were developing a more personal system using a Rockwell AIM-65 computer with a built-in printer. Their ultimate goal was to develop a portable system the same size as a portable television. The potential of the personal computer to improve people's lives lay at the heart of their research: "Throughout our work, we have tried to maintain the principal that, wherever possible, we should use the 'intelligence' of the personal computer to minimize the physical and mental effort that handicapped users must expend to use our communication aids. To the extent that children voluntarily use the EyeTracker Communication System, we will feel that we are successful in our efforts."[17] Its developers intended the EyeTracker to be quick to learn and easy to use so that people would want to use it to communicate.

The third-place winner, Robin L. Hight, developed a system for deaf people, the Lip-Reader Trainer, for the Apple II. It could convert typed text into animated mouth movements that users could study to better learn lip-reading. A teacher could use the program to construct multiple-choice tests showing mouth animations and a number of possible answers. The program created the animation from a phonetic sentence the teacher gave it using nineteen different mouth shapes. Students could view the animation as many times as needed, adjusting for animation speed. Hight did not design the system to replace face-to-face teaching of lip-reading; it was meant to be a supplement that students could use when not working with a human teacher.[18]

The seven honorable mentions of the national finalists consisted of a wide variety of inventions, again all using modified off-the-shelf technology. Joseph T. Cohn developed an Augmentative Communication Device that could use a personal computer to help severely disabled people to communicate. Users could control a variety of switches via small muscle movements (e.g., raising an eyebrow or rotating the forearm), in order to select words or pictures that the Apple II computer would display.[19] A future chairperson of SIGCAPH, Randy W. Dipner, created the Micro-Braille System, which could print braille cheaply, using modified commonplace microprocessor hardware instead of an expensive mechanical brailler. Dipner's system took text a user typed in, translated it into braille, and then printed it using a standard printer. He initially used an Intel 8080 and then switched to a Radio Shack TRS-80 Model III personal computer.[20] Similarly, Robert Stepp developed an inexpensive braille word processor. Stepp's system used an Apple II with a modified keyboard to allow the user to type in braille and edit it with a word processor.[21] Sandra J. Jackson and her co-researchers created Programs for the Learning Disabled, a software system for the TRS-80 to help teach students math and language skills. They designed it so that users would learn how the software worked by playing with it; the software in turn would alter its difficulty based on the student's performance.[22] David L. Jaffe used ultrasound technology in his Ultrasonic Head Control for a Wheelchair to allow a user to control their wheelchair via head movements. This noncontact system would track a user's head and wheelchair movements, so that the user could direct the chair, and it would automatically avoid obstacles. The system used parts from a Polaroid camera to detect the wheelchair's distance from other objects.[23] Paul F. Schwejda created the Firmware Card Training Disk, which provided a specialized interface for users to use any software on an Apple II Plus with their own adaptive devices. Schwejda's goal was to get away from providing only specialized software for computer users with disabilities and, instead, to offer a way for them to use any software. His device simulated a keyboard for the computer, so the user could plug in whatever interface device they worked with and control it, with the computer acting like the device was a standard keyboard.[24]

Finally, Raymond Kurzweil, the prominent transhumanist inventor and future author of *The Age of Intelligent Machines* and *The Age of Spiritual Machines*, developed his Reading Machine for the Blind to convert text to speech using text of any size or format. The machine possessed an unlimited English vocabulary through the use of a text-to-speech synthesizer. This was before Kurzweil was well known for his utopian futurist theories on the acceleration of scientific and technological progress, but he was already recognized during the 1980s for his inventions to aid people with disabilities. According to Kurzweil, his Reading Machine was the first computer that could convert text to synthetic speech. The device received inputted text from a scanner, used optical character recognition software to read the text, and converted it to speech.[25] Though expensive, Kurzweil's Reading Machine would become the leading consumer technology to translate text for blind people.

That these innovators could create these technologies using off-the-shelf products demonstrated some of the potential of the personal computer. Its ability to be modified into assistive technologies that it was not explicitly designed for demonstrated its value as a universal tool. Yet, for the most part, developers did not design these first personal computers considering user-friendliness or the possibilities that people with very different bodies might want or need to use the technology. Large computer companies were not yet building in standard accessibility features. Developers still considered the imagined user to be someone with a "normal" body and "normal" needs, not a multiplicity of possible differences. The first consumer-available accessible personal computer technologies followed from innovators like those in the Johns Hopkins contest, who evolved small, third-party companies creating these technologies for people with disabilities.

Accessible Personal Computer Technology for Consumers

By the mid-1980s, growing interest in the ways computers could be used by people with disabilities led to the emergence of businesses dedicated to creating assistive hardware and software. These companies were mostly tiny—many started by individuals building devices for their friends or family with disabilities—and were highly specialized, demonstrating an extreme form of niche production for a limited number of potential users. These developers created technologies that mostly allowed users with disabilities access to personal computers in their homes. At the same time, large computer companies paid little attention to building in accessibility features, while inadvertently making their machines more difficult to use for people with disabilities. Two books published in 1984 described the state of available accessible computer technology and recommended devices to work for people with different accessibility needs: Peter A. McWilliams's *Personal Computers and the Disabled* and Frank G. Bowe's *Personal Computers and Special Needs*. These types of books, part technological analysis and part buying guides for accessible computer technology, began to appear regularly in the mid-1980s. McWilliams, a journalist and author of self-help books, explained the basics of personal computers, interviewed technology developers working on accessibility, and listed details of available personal computers and the adaptive devices that worked with them. Similarly, Bowe, a prominent disability studies and special education professor and a disability rights activist, focused on current and future specialized computer devices, as well as ways general software could be utilized by people with disabilities. The authors organized their books by type of disability (e.g. vision, hearing, speech, mobility, and learning), specifying technologies appropriate to the needs associated with each. The variety of technologies these authors survey mirrors some of the diversity of user needs and demonstrates the possibilities for innovative technological development when designers take these needs into account.

For computer users with hearing impairments, personal computers offered a new method of telephone-based communication. In addition to all the other functions a computer provided, it also worked as a Telecommunications Device for the Deaf (TDD); users could call other computers through their modems and communicate directly with people on the other end by typing. Editor and journalist Henry Kisor described the effect of his personal computer on his life, in a 1984 interview:

1. For the first time, I am able to roll up as large a phone bill as my wife does.
2. For the first time, I am able to communicate with hearing people without having to look at their lips or write them letters and wait days for them to be delivered.
3. I am able to interview writers on the phone, as you are doing (though this is still a matter of potential . . . most writers still use the goose quill, not word processors, and you can forget about modems, so far as they are concerned).[26]

The computer now could provide the same functionality and social interaction as the telephone—even over the same phone lines—for deaf and hearing-impaired people. By becoming early adopters of the newest computer technology, users with disabilities found new solutions to their needs and, in turn, were important in communicating the possibilities of the technology to both disabled and nondisabled users. One author recommended portable computers, in particular, to hearing-impaired users taking advantage of early network communication capabilities, so that they could participate in phone-based communication regardless of where they were.[27]

Writing from personal experience as a deaf person, Bowe discussed computer technologies that could help deaf and hearing-impaired people learn to write and speak. Like McWilliams, he described the literacy problems hearing-impaired people faced if they lost their hearing at a young age. Bowe suggested that even a basic grammar checker in word processing software could hugely improve the writing of deaf people, though at this time such a feature was still only a distant possibility. To help deaf people and others with speech impairments, he also hoped that speech recognition software might one day provide a way for people to train and practice speaking using their computers.[28]

Unlike deaf users, blind people required more specialized adaptive devices in order to use personal computers—specifically, to understand what was displayed for users visually on the computer screen. At this time, Kurzweil's Reading Machine was the most advanced speech synthesizer available. It could read almost any text given to it, at an 80 percent accuracy rate, but it cost the enormous sum of $29,000.[29] The machine was powerful and used by those who could afford it, such as Judge Leonard Suchanek, who still found its accuracy rate unsatisfying. Then the highest-ranking employee in the federal government with multiple disabilities, Suchanek utilized multiple examples of state-of-the-art assistive technology. He used an LED-120 braille printer from Triformation Systems; it could print 120 characters per second and included Duxbury's Braille Translator software for $14,000. His office also used a portable brailler from Maryland Computer Services, the Perkins Brailler, which cost $3,000.[30] More affordable devices did exist but had less capability. The Echo II speech synthesizer from Street Electronics was popular and cost only $130; however, one author described the device's speech as sounding "like a robot with Wiener schnitzel stuck in its throat."[31] Beyond the prohibitive cost of many of these devices, speech synthesizers also were unable to translate any kind of graphics or images into speech, did not work well with spreadsheets, and were prevented by computer memory limitations from storing large numbers of individual words. Few off-the-shelf software programs were available that worked with speech synthesizers at all.[32] Blind people who wanted to use their computers with braille could buy the VersaBraille System to edit and print braille, for $6,700.[33]

For people with vision impairments but some ability to see, the fixed size of text displayed by early personal computers caused difficulties. Some specialized software existed to allow magnification of text and zooming, for example, Large Type, which allowed for typing in large-sized print, or PC-LENS, which could zoom in on portions of an IBM PC screen and worked with monochrome and color monitors.[34] These kinds of specialized technologies would eventually become unnecessary as software and operating systems adopted zooming and text size options as standard features. The move toward universal design as a more mainstream concept would make such features normal options for users in the following decade. Until there existed greater availability and easier-to-operate built-in control of computer functionality for users, however, people with disabilities required specialized technologies like these in order to use computers in the ways that worked with their individual needs.

Computer users with speech disabilities were similar to deaf users, in that their disability had little effect on their everyday computer use, but they could benefit from computer technology in terms of communication. People with speech impediments could buy the same speech synthesizers that blind people used and use them as a means of communication instead of as a way to understand computer output. Previously, people who were unable to speak had used written messages or point boards (boards with common words, phrases, and symbols that the user could point to) to communicate.[35] Similarly, portable speech synthesizers, which contained a small number of preloaded words and phrases, allowed people who could not speak to carry a speech output device with them in public. These devices were especially useful for communicating simple instructions over the phone or acquiring help during an emergency. (For example, a button could be programmed to say out loud, "I have an emergency, please send help.")[36]

Computer users with disabilities affecting their hands and fine motor control needed specialized input devices in order to enter information into a computer and manipulate software. For people who could press standard keyboard keys one at a time, third-party companies developed adaptive keyboards that circumvented multi-key presses by treating the control and shift keys like the caps lock—pressing the key once turned it on until it was pressed again to deactivate it.[37] Computer users who needed only a stabilization aid could use keyguards—simple, plastic boards that covered the keyboard and had holes for individual keys, so that a key could be depressed without neighboring keys accidentally being hit.[38] More complicated input devices were developed for people who could not use a standard keyboard at all. Users with various degrees of motor control could customize large, programmable keyboards, and those who could not use any kind of keyboard could control a personal computer via switch devices. A man with ALS wrote an essay on his disability and computer use, composing it by operating a single switch with one of his eyebrows, thus demonstrating how even the smallest muscle control could allow someone to use a personal computer.[39] Early personal computers could not interact with these kinds of input devices directly, however; the computer needed an interface card to recognize an adaptive device.[40]

People with disabilities did not only benefit from personal computer technology specialized for their specific disabilities. General technological improvements to computer hardware and software could allow for accessibility, regardless of the intended user. Bowe's advice that personal computers should be made accessible while the technology was still in its infancy, so that barriers did not have to be removed later, was beginning to be followed by the mid-1980s. Ideals of what would later be called universal design—creating technologies to work for as many users as possible through incorporating flexibility and options to accommodate needs—were slowly permeating the culture of computer developers.

Flexible technologies also held the possibility of unintended uses. Even the simple spell checker or thesaurus already built into word processors had additional uses—for instance, helping deaf people learn English.[41] General word processing software could also help people with learning disabilities who struggled with handwriting. According to Bowe, what made computer technology special compared to traditional educational tools was that it allowed for "creative learning" and versatility; it offered different ways for people to learn by being adaptable to their individual needs.[42] For example, the LOGO computer language, which operated via graphics instead of words, provided a way for people who had trouble with text to write computer programs. For people with motor disabilities, the development and expansion of Internet services provided a way to accomplish various tasks, such as shopping or banking. These services saved time for all users, but were especially beneficial for computer users who had difficulties getting around outside their houses.[43] This example demonstrates an important aspect of accessible everyday technologies: features that increase usability for all users can also act as assistive technologies for users with certain disabilities. (While a curb cut makes it easier for (almost) everyone to use the sidewalk, it is a necessity for wheelchair users.) Usability and accessibility are intertwined; designing for one can allow for the other.

Of the multitude of technologies developed by small, third-party companies during the early 1980s, many later became standard features of personal computers or a relatively common part of the consumer market for external devices. Personal computer operating systems absorbed many of these technologies, such as zooming, text enlarging, and even text-to-speech and speech-to-text functionality. Most of the small companies that developed these technologies did not last long, nor were they very successful. There were exceptions, however. Some of those companies that dealt with technologies for severely disabled people, such as whole computer systems geared toward an individual's use, or those building highly specialized devices whose complexity or limited market value made them unlikely to be taken up by larger companies found success in niche markets. For mainstream accessibility, however, it would be a number of years before such technologies standardized and became easily available.

The state of accessible computer technology in the first few years of consumer-available personal computers was replete with unactualized potential. The computer promised to change people's lives for the better, but major computer companies paid little attention to people with disabilities as computer users.

The design of personal computers was embedded with assumptions about what kinds of people would be using them, and computer technology evolved that left people with certain kinds of disabilities unable to use off-the-shelf products easily. Until the mid-to-late 1980s, accessible computer technologies were mostly developed by small companies or individuals seeking solutions to problems friends and family members encountered using the computer. Information for people not working in the computer industry or on the forefront of the technology was difficult to come by. In order for technological solutions to benefit users, both innovations *and* a social infrastructure that can disseminate them are necessary. The existence of a technology alone is insufficient to cause its uptake by users without some means by which they can become aware of it and some way to acquire it.

One way to navigate this environment—in which accessible computer technology was being slowly developed but a lack of information or organization kept it from reaching potential users—was for consumers to pool their resources and knowledge and attempt to influence the growth and use of computer technology. Disability and technology organizations, many started by parents or teachers, helped to create this bridge. They functioned as networks of information and advocacy, encouraging people with disabilities to learn about technologies that might benefit them and urging developers to address the diverse needs of their users. One such consumer-based organization was the Disabled Children's Computer Group (DCCG), a local, parent-run advocacy group in Berkeley, California. The DCCG acted as a point of convergence—a place where awareness of the possibility of technology encountered the realities of technological development, knowledge of how to use the technology, and a consumer base informed about the technology. This convergence makes the DCCG a particularly good example of the kind of social infrastructure that is necessary in order for accessible technology to reach users with disabilities.

Unicorn Engineering and the Brands

The Disabled Children's Computer Group evolved out of a connection with a small, third-party, accessible technology developer. Unicorn Engineering, a company making adaptive keyboards for people with motor disabilities, was typical of the small businesses developing accessible personal computer technology in the early years. The company, and the man who started it, would go on to play an integral role in the founding of the DCCG. In 1979, Steve Gensler, a resident of Oakland, California, started Unicorn Engineering after creating a computer keyboard that could be used by a friend with cerebral palsy. This kind of origin—one technologically proficient person trying to making a computer work for a friend or family member with disabilities—was common among accessible computer technology start-up companies. Gensler taught himself electronics in order to build a keyboard with large, flat buttons that were easy to press and were programmable, so that the computer could be instructed to respond to any key in any manner desired. In the 1992 patent application for the successor to this keyboard, engineers from IntelliTools (the company that Unicorn Engineering became in 1991) described what made the Unicorn Board particularly usable by people with disabilities.[44] The patent compares a traditional keyboard to the Unicorn Board; the former requires roughly the same level of dexterity as operating a typewriter, whereas the latter can have "keys" of any desired size and configuration and assigned for any function, making it operable by people with varying degrees of motor control. Instead of the individual keys found in a standard keyboard, the Unicorn Board had a flexible membrane covering hundreds of switches (fig. 2.1). The membrane could be divided into any number of programmed sections, each covering a number of switches.[45] Each section, or "key," would be labeled by a card that covered the entire membrane. As the labels on the overlay were customizable, users could display whatever symbols, colors, numbers, or words suited their needs.[46] When a user pressed anywhere on the overlay within an area assigned to some function, the flexible membrane would activate the switches underneath, and the computer would read the switches. This level of customization allowed the Unicorn Board to meet the needs of any user able to press a button of any size.

In 1984, rehabilitation researchers at a Closing the Gap conference described their use of the Unicorn Model 1 Keyboard for their clients with disabilities.[47] Their Unicorn Board could be programmed to have up to 128 keys that performed different functions. Because the keyboard was this flexible, it could be programmed to optimize the user's interaction with it in operating specific software. That is, only those keys needed to control the desired software had to be used, and they could be the size the user could best operate; no extraneous keys would be present that the user might accidentally press.[48] Additionally, the researchers found it advantageous that the most frequently used keys could be programmed to be those easiest to reach.[49] As with other accessible interface devices at this time, however, the membrane keyboards developed by Unicorn could not communicate directly with the computer; they required a separate interface card that could make the computer understand the input from the keyboard. The interface card translated a press of the keyboard into information the computer could read; the computer would then think that it was reading input from a standard keyboard.

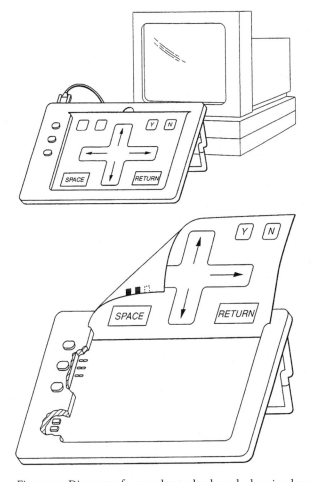

Figure 2.1. Diagram of a membrane keyboard, showing key overlay on top of switches. The overlay in this example is divided into large, simple key commands, such as arrows, space, return, yes, and no. From the 1992 Intellitools patent application for the successor to the Unicorn Keyboard.

The benefits of the Unicorn Board for people with certain disabilities brought together Steve Gensler and Jackie and Steve Brand, two teachers in the San Francisco Bay area. Gensler met Steve Brand when they both took computer classes to learn more about the technology that might help the people they cared about. Gensler sought to learn enough to build adaptive keyboards, and Brand was looking for solutions that would enable his daughter Shoshana to use a personal computer.[50] (Gensler's Unicorn Board became one of the first technologies to offer the Brand's daughter a way to help her communicate and learn.[51] This initial, very personal connection with computer technology showed the Brands the liberatory potential of personal computers.) Due to complications during her birth in 1974, Shoshana had developed cerebral palsy and vision impairments. As a consequence, she was unable to speak, did not possess fine motor control, and was legally blind. When Shoshana was still a baby, the Brands looked for programs and services in the area that could help them with the difficulties they faced raising a child with disabilities. Jackie Brand sought help from the Center for Independent Living (CIL) in Berkeley. Although the CIL focused on adults with disabilities and gave little attention to parents of children with disabilities, she went to work there to gain what help from and familiarity with the disability movement that she could. One of the people who inspired Jackie Brand at CIL was Judy Heumann, a prominent leader of the disability rights movement. Heumann influenced Brand with her views on the need for independence for people with disabilities and persuaded her not to be overprotective of Shoshana, who needed to be given the opportunity to live her own life. This view of people with disabilities living full, independent lives and participating in society as they desired would shape the disability and technology organizations Jackie Brand would go on to create. Seeking ways for Shoshana to live an independent life led the Brands to explore accessible computer technology as a tool that could benefit their daughter. The promise that technology might enable people with disabilities to live in society fully, combined with the complicated and confusing state of accessible computer technology during the late 1970s and early 1980s, pushed the Brands to find their own solutions for their daughter and to join together with other dedicated individuals looking for answers.[52]

Jackie Brand's first interaction with other parents of children with disabilities was through a local parent support group at the Alameda County Association for the Mentally Retarded. She found, however, that few other parents thought about their children in the long term—as children with disabilities who would grow up to be adults with disabilities. This lack of forethought was strikingly apparent during the April 1977 San Francisco sit-in at the Department of Health, Education, and Welfare offices to protest the delay of the signing of the Section 504 regulations. Jackie Brand went to the federal building to join the sit-in with three other parents to whom she was close. Heumann asked them to try to gather a group of parents and their children with disabilities to protest in front of the building. Calling all the local parents she knew, Brand found that most parents were unreceptive, claiming that this protest was not relevant to them because their children would not grow up to need the protections of legislation such as Section 504. She saw in these parents' position a refusal to accept the reality of their children's futures; they thought that their children would not need civil rights protections because they could not imagine them living the kinds of independent adult lives that the Brands and the CIL desired for people with disabilities. This lack of perspective limited the usefulness of parent support groups for Jackie; she saw her own views about her daughter's future as too different. The Brands stayed close to the few parents of like minds, though, and this handful of parents desiring independence for their children became the initial group that formed the DCCG six years later. Their disability and technology advocacy organization would help inspire the creation of social technologies that would connect thousands of disabled people with beneficial computer technology.[53]

Other legislation in the 1970s, however, more immediately affected the parents of children with disabilities. The Education for All Handicapped Children Act of 1975—the beginning of mainstreaming of education for children with disabilities— brought disabled and nondisabled children together into the same classrooms to receive the same education. Jackie Brand worked on a CIL offshoot project called Keys to Introducing Disability in Schools, which, with the advice of children and adults with disabilities, developed curriculum that would be used in classrooms with both disabled and nondisabled children. The goal of the project was to make children and teachers comfortable in integrated classrooms. Integrating children with disabilities into regular classrooms was not easy, however, and the Brands ran into problems trying to find a school environment that fit Shoshana's needs. They struggled at a number of different schools, including a mainstreaming elementary school in Berkeley where their daughter was ignored and not taught the same materials as other children, a special education program in the Richmond public schools where Shoshana was taught only with other children with disabilities but interacted with nondisabled children elsewhere in the school, and a Richmond middle school with a relatively successful, fully mainstreamed program that unfortunately lacked a necessary elevator to have physically accessible classrooms.[54]

Accessible buildings were not the only requirement for Shoshana to be able to learn. The need for other kinds of assistive technology introduced Jackie and Steve to personal computer technology and got them thinking about how to make it usable by people with disabilities. The Brands wanted their daughter to be a part of the computer revolution they were witnessing in the early 1980s. They hoped that computers could allow her to learn and communicate in ways that Jackie found other tools did not allow.

> For example, books: she couldn't read the books. For example, the blackboard: she couldn't see the print on the blackboard. For example, pencils and paper: she didn't have the fine motor capacity to write, and so though she had a great interest in the academics of school, the tools weren't right for her. The tools didn't work for her, and we didn't know exactly what would work for her, but here was this, you know, new microcomputer revolution before us with promises that it was going to just revolutionize learning and teaching and the way in which we function in society. And our first thought was, well, why shouldn't these revolutionary new tools work for [Shoshana] in place of some of the tools that were in school that weren't working for her.[55]

The computer opened up possibilities for the Brands that other tools had not allowed (fig. 2.2). By adapting to Shoshana's own needs and accommodating her disabilities, she could learn and communicate in ways she was otherwise prevented from doing. First though, Jackie and Steve Brand needed to learn about personal computer technology themselves. Steve took a year of leave from teaching in order to take classes and learn enough about computer programming and computer hardware to put together a system that Shoshana could use. It was during this time that the Brands met Steve Gensler and were introduced to his fledgling Unicorn Board.

Gensler gave the Brands a very early version of the keyboard to see if they could get it working properly with the Apple II computer they were setting up for Shoshana.[56] Using an interface card that allowed the computer to recognize the keyboard, Steve Brand was able to program the Unicorn Board. One reason the Brands found it so useful for Shoshana was that it was programmable and could be adapted to her needs. The keyboard allowed Jackie and Steve to assess what degree of vision and hand control their daughter possessed. They began by having the computer respond positively—for example, by playing music—to Shoshana touching anywhere on the keyboard at all, as a way to show her that she could control simple cause and effect.[57] The Brands then added colors that they knew Shoshana was able to identify, covering each half of the keyboard overlay with red or yellow and asking her to touch one side or the other. They then divided the keyboard into four colors or pictures of different animals and asked her to press each by name. The computer responded to the animals, for example, by making the cor-

Figure 2.2. Shoshanna Brand, in 1983, struggling to use a pen and paper. As a child, she found that the fine motor control required to use writing tools effectively presented an obstacle to her. Courtesy of Jackie and Shoshanna Brand.

responding noise each animal makes. As Shoshana quickly learned how to use the computer, her parents programmed the Unicorn keyboard to allow her to communicate via text, by having buttons represent sentences of words.

This feature of ever increasing complexity allowed Shoshana to learn how to use the computer with the Unicorn Board (fig. 2.3). Jackie Brand describes the way her daughter learned: "Eventually the keys got smaller and smaller, there were more and more divisions on that board, until she had essentially a full keyboard to work with. Had we shown her that full keyboard right at the beginning, there was no

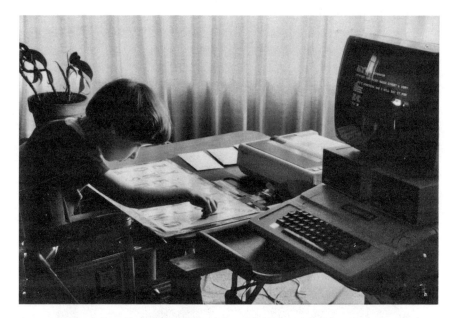

Figure 2.3. Shoshanna Brand using a mock-up of the Unicorn Keyboard, demonstrating the greater ease she had using this technology rather than traditional communication tools. Courtesy of Jackie and Shoshanna Brand.

way she could have done it. She needed to build her ability to distinguish and to move between smaller and smaller distances. That was the developmental thing she needed, and this keyboard uniquely provided that opportunity."[58] The Unicorn keyboard was powerful because it was adaptable. People with many kinds of disabilities that affected how they might operate a computer keyboard were all able to use this technology by programming it for their individual skill levels. Such a device could make the computer more truly universal by working for people with different kinds of abilities and needs.

However, as great as the potential of assistive technology was to bring the personal computer to people with disabilities, Jackie Brand also realized that these devices were not yet usable for a large audience. The Unicorn Board and other adaptive input devices could not simply be plugged into a computer and work correctly; they required a complicated setup through an interface card that translated the input device for the computer: "This keyboard and interface card did a lot of things for a lot of kids and adults who otherwise were really blocked from accessing a computer, so this was the beginning of my understanding about the power of assistive technology. And we also realized that this was not easy stuff to do. It would have to be a lot easier to use before many people would benefit from it."[59] Accessible computer technology was far from simple or intuitive in the early 1980s; using it required a steep learning curve and multiple devices working together. The Brands found that the computer industry and the technology itself was not yet set up to make accessible technologies readily available. Small companies and entrepreneurs developed adaptive devices that allowed people with various disabilities to use computers, but information on such devices was not readily available, and making different interface devices work with early personal computers was not easy. Large computer companies, such as IBM or Apple Computer, produced new innovations quickly as the personal computer developed but, until the mid-1980s, were not explicitly focused on addressing the needs of people with disabilities.

Dolores Hagen, cofounder of Closing the Gap—a conference and journal on technology in special education—argued that a lack of information was a barrier preventing personal computer technology from reaching more people with disabilities during the early 1980s. She described a rhetoric surrounding personal computers that made the technology sound intimidating and complicated. Misinformation led potential users to believe they needed to have skills in programming or math in order to operate the computer. Marketing by computer companies declaring that anyone could learn to be a programmer turned away people who might otherwise have been interested in computers but believed that this perceived requirement to learn programming put the technology beyond their reach.[60] The hobbyist origins of the personal computer in this way worked against its quicker adoption by the general public. Apple, for example, tried to counter the view that the computer required specialized knowledge to use by featuring an everyday object as their logo. Their advertising slogan for the Apple II was "Simplicity is the ultimate sophistication."[61] By 1984, Hagen believed the situation was improving, but she still saw a lack of communication between programmers and teachers, in particular, regarding the needs of special education students and the use of computer technology.[62]

In spite of these difficulties, however, the Brands slowly found personal computer equipment for Shoshana. They discovered that the computer empowered their daughter to be able to express herself in ways people had always told them she would never be able to, and it changed the expectations both the Brands and outsiders had of her. As Jackie remarked, "Imagine the whole creative process opening up before you, where there has never been an outlet before. Imagine suddenly being able to express your needs, your desires, when you've never spoken or written a word. That is a drama that makes chills go up your spine."[63] The computer made it clear that, although Shoshana could not use traditional means like speech or handwriting, she could in fact communicate; the technology behind her communication just needed to accommodate her abilities. Alongside the Unicorn Board, Shoshana also used a speech synthesizer, the Echo II, from Street Electronics.[64] For the first time, she was able to write out what she wanted to say and use the computer to speak it out loud to others. The Echo II also allowed Shoshana to work more independently, as it could accommodate her vision impairment by reading aloud any text she composed, whenever she wanted it to, allowing her to edit her work herself.[65] Brand described the way these computer technologies made her and her husband rethink what their daughter was capable of: "And all of a sudden, 'She can't do this,' became, 'Wait a minute. We haven't found the tool to help her do it yet.' And so it was [an] enormous mental change and shift in attitude to begin to understand just the very tip of the iceberg of what technology might mean."[66] For the Brands, computer technology became the tool that would enable their daughter to communicate and learn at her full potential. Shoshana grew up in the midst of steadily evolving computer technology; she would adapt it to her needs, as she

Figure 2.4. Shoshanna Brand, age twenty-one, using a more advanced computer keyboard from Intellitools. The adaptability of her first Unicorn Keyboard enabled her to learn to use increasingly complex devices. At the same time, the technology itself improved to better meet the needs of its users. Courtesy of Jackie and Shoshanna Brand.

learned to use continually more complex and flexible technology (fig. 2.4). The value embedded in the personal computer as a universal tool made it precisely the right tool to be adapted to individual needs.

The Disabled Children's Computer Group

Desiring to share what they had learned of the possibilities of computer technology for people with disabilities and to combine their efforts with others,' Jackie and Steve Brand joined with a handful of local parents to found the Disabled Children's Computer Group in November 1983. The group held meetings and technology demonstrations at the University of California, Berkeley, using donated space in the Lawrence Hall of Science, a public science education center. At their first official meeting, this small group of parents were joined by adults with disabilities, teachers, medical professionals, and people working in technology fields— around fifty people total.[67] According to Jackie Brand, the DCCG's meeting location played a role in the organization's quick growth: "The Lawrence Hall of Science had a history of providing hands-on science experiences for children, including computer awareness days. In addition to its programming expertise, the Lawrence Hall of Science provided two essentials for any new organization: a place to exist and a way to do mailings. To this day, people active in the DCCG feel that this original alliance between interested parents and an established professional organization has been key to success."[68] The Lawrence Hall of Science offered the right space for the DCCG, providing a part of the social technology needed to create a network of knowledge transmission about computer technologies for people with disabilities. That the hall met the group's needs so well also demonstrates the importance of geography and the role played by Berkeley itself in developing such social technologies. At the university and around the city, Berkeley's culture of activism and its history in the struggle for civil rights helped an organization like the DCCG find the support and encouragement it needed to grow. Berkeley was the center of the disability rights movement in the 1970s; the same students with disabilities who fought for access to UC Berkeley went on to create the Center for Independent Living and inspire Jackie Brand's views on disability and independence. People who worked for the Lawrence Hall of Science, such as Linda DeLucchi and Larry Malone, were among the first members of the DCCG and helped it grow, organize, and find funding.[69]

The initial connection between consumers (parents of children with disabilities and adults with disabilities) and professionals (medical and technology professionals) at the group's first meeting became one of the major values of the DCCG and continues to the present day. A spokesperson for the DCCG described the reasoning behind this value of combined effort: "Parents bring an urgency and commitment which is complemented by professionals' experience and resources."[70] The very personal passion of parents—wanting their children to enjoy the same benefits of computer technology as other children—combined with the interests of technology developers—many of whom were motivated by close relationships to people with disabilities or were disabled themselves. Together these groups worked to promote accessible technologies that brought people with disabilities in as users and consumers of computer technology. Steve Brand described the parents' perspective two years after forming the DCCG (fig. 2.5): "We saw the power

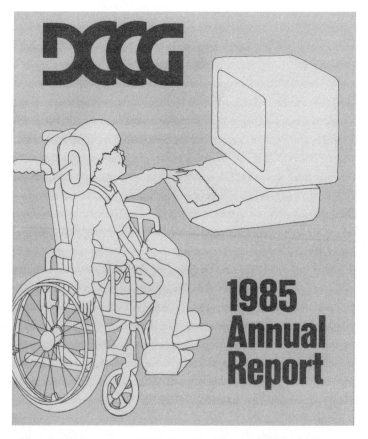

Figure 2.5. Cover of *1985 DCCG Annual Report*, emphasizing the role of disabled children as computer users within the organization. Courtesy of Jackie Brand.

that the computer could have for our children, and we wanted all the current applications to be available to them."[71] By working directly with computer professionals, these parents created a network of information sharing and feedback that helped provide their children with technology they could use; this core structure of partnership was built into the foundation of the DCCG and represented the values of the people who ran the organization. The relationship between consumers and professionals framed the DCCG's work in connecting disabled users with computer technology that could benefit them and in the formation a few years later of the National Special Education Alliance from within Apple Computer.

The DCCG grew initially through word of mouth. At its creation and for the next few years, Jackie Brand worked as the group's executive director and Steve Brand as its president. The group established a steering committee after the first meeting to plan future meetings and workshops; they would become the initial board of directors. During the first two years, the DCCG held around five general-purpose meetings per year, where members and other interested people could meet and share their knowledge and resources. Those knowledgeable about accessible computer technology were invited to give demonstrations showing the strengths and weaknesses of devices or software.[72] The group also held occasional weekend hands-on workshops, where parents and potential users of computer technology could interact with devices directly to test them out and see what worked best for their individual circumstances.[73] This network of expertise sharing permitted people of various backgrounds to become experts in the burgeoning technology and to share what they had learned with others. Clearly addressing needs in the community beyond only a small group of parents, the DCCG grew quickly. Within two years, the group had one thousand members. At this point, the organization was so large that its needs could no longer be served by the Lawrence Hall of Science. The DCCG needed a more permanent home.

In September 1985, a local elementary school donated spare classroom space to the DCCG; here, they started their resource center, a semipermanent facility for holding meetings and storing computer technology. At the resource center, the group was able to expand the services and activities they offered. The DCCG continued their technology demonstrations, now conducting regular, personal, hands-on consultations for parents, adults with disabilities, and disability professionals. Open meetings still took place where professionals and community members with varying levels of computer knowledge discussed technology. The group also conducted presentations for other organizations, including special education groups, family support groups, and computer clubs, such as the California Educators for the Physically Handicapped, the Oakland Association of Chinese Parents of the Disabled, and the Homebrew Computer Club. Making use of increasingly available online computer technologies, the DCCG ran an electronic bulletin board to better disseminate new information on technologies and events. This kind of telecommunications technology would go on to be an important component of the National Special Education Alliance a few years later. The space allocated to the DCCG at the resource center also allowed the group to maintain their own collection of computer hardware and software that could be demonstrated and loaned out to members.[74]

One of the purposes of the resource center was to demystify the confusing world of computer technologies for parents who lacked adequate information. The DCCG held an annual workshop where families could learn about available accessible computer technologies and discover what worked best for them. Jackie Brand described serving this need for local families: "As many people will testify, it is difficult enough to get clear, understandable answers to questions about how and what to buy when there are no specialized needs or adaptations necessary. When there are, and one is uncertain about what might work, there is almost no commercial establishment able to give adequate information about special education needs. For those families who think their daughter or son might benefit from a computer, but do not know which kind or what adaptations are necessary, the DCCG Family Workshop offers an important first step."[75] The DCCG thus acted as a social technology for consumers.

The family workshops also provided an opportunity for children with disabilities to be the experts, sharing their computer experiences with other families. Brand explained that this was an uncommon role reversal for many children with disabilities, as they were usually the ones receiving help. She described the variety of technologies these children showed off, "including large print screens, braille embossers, software for talking Apple Computers, left hand only keyboards, single switch computer controls, computers for non vocal individuals, and robots."[76] By sharing their knowledge of these cutting-edge technologies, these children demonstrated not only that were they capable of mastering complicated technology but that they could teach its use to others (fig. 2.6). The consumer/professional partnership behind the DCCG also played a role in the family workshops; technology developers would bring their products to demonstrate and allow users to try out. This was not a one-way exchange, however, as consumers also provided feedback to developers about what was working for them, what could be improved, and which of their needs were not yet being met.[77] There was an environment of prob-

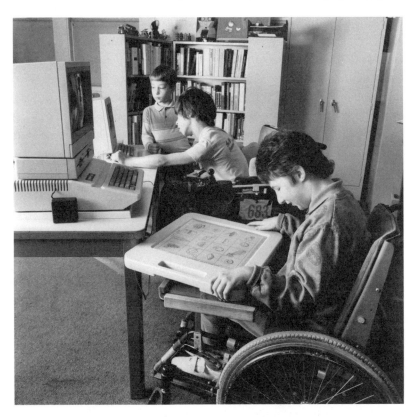

Figure 2.6. Children using accessible computer technology at the DCCG in 1987. The DCCG provided a place for children with disabilities to interact with and teach one another about cutting-edge computer technologies. Courtesy of Jackie Brand.

lem solving reminiscent of the kinds of hobbyist tinkering with personal computers that had helped start the personal computer industry roughly a decade earlier.

While the DCCG quickly grew during the mid-1980s, Unicorn Engineering also found success, partly in connection with the DCCG. During the winter of 1984–85, Arjan Khalsa, a teacher in the Bay Area, became interested in accessible computer technology. Khalsa had recently attended a class on mainstreaming special education students in the Berkeley area and was struck by the role technology played in accommodating disabilities, as well as the obstacles technology created when it was not designed to be accessible. Shortly thereafter, Khalsa happened to catch a radio program discussing McWilliams's book on computers and people with disabilities, and he contacted the author for more information. McWilliams directed Khalsa to the DCCG, where he met Gensler. Moved by the meeting he attended, Khalsa threw himself into working with disabilities and computer technology.[78] Khalsa soon partnered with Gensler, eventually becoming the CEO of IntelliTools after it replaced Unicorn Engineering. Khalsa also became a long-standing member of the DCCG's board of directors, for years helping to shape the direction of the organization. Unicorn Engineering also continued its ties with the DCCG; into the 1990s, Unicorn Board users held regular meetings at the DCCG and conducted seminars for people new to the technology.[79] The DCCG provided a space where developers and users could directly interact and share their knowledge with others.

Computer Companies and Accessibility in the 1980s

The lasting success of Unicorn Engineering was not typical. While there were many small companies making accessible computer technologies in the mid-1980s, most were unable to make enough money to sustain themselves. Jackie Brand describes the situation: "Most of those companies didn't last, and so a great solution would be around but they didn't have the resources to market those solutions, to produce those solutions in enough quantity to bring prices down, and so they were very expensive solutions that very, very few could afford. Eventually most of these companies just went out of business."[80] Companies such as Unicorn Engineering were the exception in finding success while remaining fairly small.[81] The failure of most of these niche companies demonstrates one of the problems with relying on market solutions to the social problem of disability: companies that could not find success in the market would be unable to make products that might benefit people, continuing the problem of lack of access.

What small companies did have that large computer companies lacked was the enthusiasm to develop accessible computer technologies. Through the mid-1980s, large computer companies showed little interest in developing their own accessible technologies. Take, for example, Ken Yankelevitz, an aerospace engineer, who built accessible game controllers for Atari computers so that his quadriplegic friends could play games. Yankelevitz took his controllers directly to Atari in the hopes that the company would be interested in selling them. Atari declined, instead opting to refer any inquiries about accessible game controllers to Yankelevitz's KY Enterprises.[82] Because they did not see a very large viable market for accessible technologies, it made more financial sense to companies like Atari to allow third parties to create devices and software to work with their technologies.

Slowly, large companies began to create their own programs focused on accessibility in the mid to late-1980s, though it would still be a number of years before they would do so to appeal directly to the disability market, rather than out of charity or goodwill. One reason for this gradual shift was the growing attention paid to the issue by disability professionals, academics, and the government. The U.S. Department of Education put together a meeting at the White House in February 1984 for computer company representatives and disability professionals to discuss how personal computers could be made accessible for more people with disabilities. The University of Wisconsin–Madison's Trace Center helped coordinate the meeting and published the results. The goal of this meeting was to inform computer companies about issues of accessibility and recruit their support to address the problems people with disabilities faced in using personal computers. The task force provided suggestions to computer companies that they determined would be reasonable to implement and that would make public computers usable by more people. The main conclusion of the group was that access points should be made readily available; that is, public computers should have accessible ports where a user can plug in their adaptive devices.[83] Users would carry whatever specific technology they needed with them and could then plug it into any computer to use it. By building in such a degree of flexibility, any computer could accommodate many different embodied uses. This concept demonstrates the growing value of designing technology to be universally usable; however, it still required users to carry their own adaptive devices with them instead of making the computer accommodate different uses directly. This initial meeting began a formal dialogue between the computer industry and disability professionals, and in subsequent yearly meetings, these professionals continued transmitting the knowledge necessary to make developers aware of the different needs of their users and to encourage them to accommodate those needs as best as possible. Following from this increased attention toward computer users with disabilities, two companies with distinct histories and corporate values, IBM and Apple Computer, worked to create in-house accessibility initiatives.

IBM AND PERSONAL COMPUTER ACCESSIBILITY

Through the middle of the twentieth century, rival computer companies saw IBM as holding a monopoly on the computer technology industry and thereby controlling innovation. The company had a long history of production strategies that demanded that all components of a given system were produced in-house in order to stifle competition.[84] While this way of doing things would change dramatically with the development of the personal computer, it influenced how the company tackled accessibility issues. Accessibility at IBM was addressed in-house and focused on its own employees with disabilities. These efforts sometimes replicated existing accessible technologies from third-party companies, but they also involved new innovations. IBM's computer industry domination led to a Justice Department antitrust lawsuit in 1969. Twelve years later, the lawsuit was dropped, as the advent of the personal computer had changed the industry to such an extent that IBM could no longer be seen to be impeding competition, and the accusation of a monopoly was found to be "without merit."[85] Compared to Apple Computer, IBM entered the personal computer business relatively late. It was not until 1981 that the company sold its first personal computer—using the abbreviation of personal computer (PC) as the name for its product. Due to the company's power in the computer industry, the IBM PC became an instant success among business workers, far beyond what IBM had anticipated. During the years after its release, fifty million computers had been installed using some version of the PC architecture and running MS-DOS.[86] IBM became the "Big Brother" that companies created from within the counterculture, such as Apple, fought against; Apple demonstrated this stance explicitly in its "1984" Macintosh TV ad. For Apple, IBM represented an old-fashioned way of viewing computers that needed to change, one in which the computer was cold, inhuman, and impersonal—the opposite of a convivial technology that had the possibility to be a tool for any imagined use.

Despite seeming untoppable, IBM made certain decisions in how it produced the PC that led to its losing dominance of the personal computer market. The company rushed to manufacture the PC in a far shorter time than it usually developed products, beating the time crunch by incorporating parts manufactured by other companies instead of developing all aspects of the hardware and software internally, as it had with previous computers. This move created an environment where IBM-compatible computers could come into being. Third-party companies produced machines from reverse-engineered IBM specifications and followed IBM's own standards so that they could run the same hardware peripherals and software that IBM designed the PC to use. The compatibles not only were cheaper than IBM's computers but also were so easily available that the market became saturated with many companies selling similar machines. In addition, Microsoft retained the right to sell MS-DOS to companies other than IBM, allowing its operating system to quickly flood the market. IBM-compatible personal computers soon occupied the vast majority of the personal computer market. By 1991, IBM only controlled a small part of the market that it had created, and companies such as Compaq and Dell made more of a profit with their compatibles than IBM itself.[87] For the first time in the company's history, IBM experienced annual losses for three years during the early 1990s.[88] In 1987, IBM released the successor to the PC, the Personal System/2 (PS/2); it was IBM's first personal computer with a graphical user interface, running either Operating System/2 (OS/2)—a collaboration between IBM and Microsoft—or Microsoft Windows. The PC and the PS/2 were IBM's main personal computer products during the 1980s and 1990s, and these were where the company's accessibility efforts were directed during this time.

With their first personal computers released in the 1980s, IBM began to work toward the development of products that considered accessibility for people with disabilities. It developed its own accessibility products and features internally, in addition to setting up means to communicate directly with users with disabilities and activist groups representing them. IBM organized its efforts through three programs within the company: the National Support Center for Persons with Disabilities, Special Needs Programs, and Special Needs Systems. Each group was located in a different part of the United States and focused on different, though overlapping, aspects of developing accessible computer technologies and getting them into the hands of people with disabilities. This division allowed IBM to target their efforts in three areas: communicating with users, organizing efforts to develop accessibility, and creating accessible products.

IBM founded its National Support Center for Persons with Disabilities in 1986, in Atlanta, Georgia. The center's main activity was to connect people with disabilities with computer technology that could benefit them. The National Support Center communicated with medical professionals, disability agencies, employers of people with disabilities, educators, and individuals with disabilities. Similar to the DCCG's Resource Center, IBM had an equipment demonstration area where people could try out new technologies. IBM would not prescribe specific third-party technologies to people but provided information on what technologies existed and how people could acquire them. National Support Center employees presented at national conferences in order to promote the development of accessible technologies. The center also ran an electronic bulletin board that offered information on IBM-compatible technologies, hosted special interest group forums for users to communicate with each other, and provided various files and software for download.[89] In addition to these methods of connecting people with various technologies, the center provided marketing and technical support for three IBM products developed later: the Screen Reader, the PhoneCommunicator, and the Speech-Viewer.[90] Part of a massive corporation, the center was utilized by thousands of people with disabilities looking for the information and services it offered. It maintained a database that included 500 third-party technology vendors and 850 external disability support groups. In 1989 alone, the National Support Center handled more than 24,000 requests from people with disabilities and interested parties.[91]

The Special Needs Programs, out of Somers, New York, managed research projects and reviewed the design of IBM products for their accessibility features. The group began in an early form in 1979 to fund research and manage development projects that would lead to consumer products for people with disabilities.[92] Bob Bettendorf, the coordinator for Special Needs Programs, explained in a 1988 *Think* article why IBM's organizational divisions were necessary in order to better develop technologies for people with disabilities. He argued that developers might come up with good ideas for new technologies, but without connections with marketing, these projects would fail, as they had in the past, either because of a lack of fit with users' needs or because of an inability to sell the product to users who could benefit from it. By coordinating and managing efforts through Special Needs Programs, IBM hoped to avoid repeating such failures. According to Bettendorf, "We're taking the enthusiasm of employees who are personally committed and coupling that enthusiasm with appropriate development and business disciplines. Our goal is to have a steady stream of successful products for people with disabilities."[93] Once Special Needs Programs found worthwhile research projects, the ideas could then be developed into products that would be both salable and usable.

Founded in December 1986, the Special Needs Systems group ran out of Boca Raton, Florida, and oversaw the development of IBM computer products intended for people with disabilities. The line of products developed through the Special Needs Systems group during the late 1980s and early 1990s was called the Independence Series. The goal of the Independence Series was "to enhance the quality of life and employability of persons with disabilities through the use of IBM technology."[94] By 1996, there were ten products in the series, including IBM's popular Screen Reader, as well as programs to add accessibility features to DOS, allow phone communication for deaf people, aid therapists working with people with cognitive disabilities, and provide speech controls for a personal computer.

IBM's organizational structure allowed the company to focus its efforts independently and cooperatively on three different approaches to personal computer accessibility. In addition to encouraging the creation of accessible technologies and marketing them to consumers, IBM also developed its own network of ways to interact with users. While the company was not an advocacy organization, it did understand that its own employees with disabilities needed technologies that would allow them to use IBM computers and that public consumers could also benefit from these same technologies. Thus, in the late 1980s and early 1990s, the company took accessible personal computer technologies initially created for its employees and marketed them as consumer products.

Apple Computer's Office of Special Education and Rehabilitation

Like IBM, Apple Computer dedicated resources to promoting personal computer technology for people with disabilities during the 1980s. However, as a far smaller company that emerged from within the activist and hobbyist milieu of the Bay Area, Apple approached computer accessibility in a more localized fashion. Steve Jobs and Stephen Wozniak founded Apple Computer in 1976, with the creation and sale of the Apple II. The company was built out of counterculture values and hobbyist tinkering. Prior to 1985 and the creation of its own internal disability group, though, Apple did not explicitly respond to the needs of their users with disabilities. While engaged in a quest to design personal computers to be more friendly and usable for everyone, Apple also made design decisions that inadvertently created obstacles for people with certain kinds of disabilities.

The Apple II, built from a brilliant and minimalist design by Wozniak, was the first highly successful complete personal computer (able to be used off-the-shelf) available to consumers. Its open design and ease of adding peripherals made it popular with those looking to make personal computer technology work for people with disabilities, even though it was still difficult to use in many ways. With the plethora of personal computers available by the mid-1980s, consumers bought whichever machine would run the software they needed.[95] Many third-party software developers wrote their programs for the Apple II because it was flexible and lacked interface standards. The result was that consumers could choose between a large number of software applications; some were user-friendly, others were not, but all had different features and were operated in different ways.[96] Therefore, because so many software options existed, the Apple II became the computer most widely used by people with disabilities for some time.

The Apple Macintosh, released in 1984, changed the experiences of all users but especially those with disabilities. The Macintosh brought a new personal computer paradigm to consumers—one in which every application looked the same and was controlled the same way, with the same menus and keyboard shortcuts. This standardization made the Macintosh more reliable to use—a user would not have to learn an entirely new interface for each program—but third-party developers had to follow the Macintosh interface rules in order to write software for it.[97] The Macintosh began with the vision of a man who wanted to enable users to do whatever they might desire with the personal computer. Jef Raskin, the initial head of the Macintosh project (which he named after his favorite type of apple), sought to create a computer with an interface designed to be easy and intuitive to use.[98] By 1981, clashes within the company drove Raskin out of Apple, but some of his vision of a truly friendly computer made it into the Macintosh.[99] Frank Bowe predicted that the ability of the Macintosh operating system to allow multiple programs to run at the same time would be particularly useful. Older personal computers could only run one program at a time, which created usability problems when multiple functions were not available in the same program (e.g., one might need to calculate the solution to a math problem to insert in a document one was writing). People with disabilities were particularly affected by the difficulty and length of time it took to close one program, insert another disk, open up another program, run whatever was desired, then return to the first program. The Macintosh allowed concurrently running programs to be interrupted and easily switched between via the keyboard or mouse.[100] This specific operating system design would benefit all computer users and demonstrated the user-friendly perspective on design and technology to which Apple aspired. This kind of increase in usability—by increasing flexibility for the user—allowed the personal computer to become more accessible for people who needed to use it in different ways.

However, not all interface changes brought about by the Macintosh benefited all users. The ideal of personal computers enabling people carried with it a certain image of who its users might be, and a number of Macintosh design decisions made using it simpler for some people while restricting options for people with certain disabilities. For example, as it was initially released, the Macintosh lacked cursor (arrow) keys on the keyboard. Apple designed it this way in order to force computer users to adjust to navigating with the mouse. Macintosh marketing representative Joanna Hoffman explained that the lack of cursor keys was also a way to force third-party software developers to create new applications for the Macintosh, which needed the mouse, instead of adapting old ones, which would have used cursor keys.[101]

This kind of design decision restricted the options available to the users, requiring them to learn a new technology Apple deemed better for them. The company feared that users would not latch onto such a large change as the mouse and would instead fall back on what they were comfortable with. Apple had much at stake in the success of the mouse—it was one of the most visible innovations that made the

Macintosh stand out—and believed that it was a superior input device for the computer. For most users, the mouse likely was the best input device; it can be as intuitive to move the cursor with the mouse as it is to use a steering wheel to drive a car. However, for anyone who found pushing a key easier than controlling a mouse, an off-the-shelf Macintosh would come with new impediments. Apple later relaxed these restrictions on user navigation choice; the Macintosh Plus, released two years after the original, returned the cursor keys to the keyboard.[102] In this case, by not imagining people with disabilities as possible users, Apple decreased flexibility and made the Macintosh less usable for people who needed different ways to operate the computer. Apple attempted to increase user-friendliness by making the computer simpler to use, with fewer options. However, trying to make the computer work for all users would involve providing for various needs by making the technology more flexible, though more complex.

While Apple stumbled over accessibility issues in some aspects of their computer development, the company also made a concerted effort in the mid-1980s to address the needs of people with disabilities by creating its own internal group dedicated to such issues. In July 1985, Alan Brightman started the Office of Special Education and Rehabilitation (OSER) at Apple Computer.[103] A former disability activist in the Boston area, Brightman had visited Apple in 1984, invited by a friend of his who worked there. Apple offered Brightman a job within its Education Foundation, a group dedicated to giving grants to schools in need. After working there for a while, he became frustrated by the fact that Apple marketed itself as caring about individuals yet lacked any explicit focus on people with disabilities. Brightman sent a short proposal to Steve Jobs and John Sculley, to which they responded positively. In a 2008 interview, Brightman described his meeting with Sculley: "[Sculley] said, 'I don't know if you're going to be able to pull this off or not, but you have to make me one promise.' He went on to say . . . 'you have to promise me that if you fail at this, you will fail huge.' And to this day, that was the best permission I ever got to go for something and that was the beauty of Apple at that time. Apple wasn't about making little waves; it was about making big waves and John didn't want it to be any different in this domain as well."[104] Sculley's charge to Brightman to try something huge in order to better include users who were being ignored harkened back to the values Apple was founded on—it sought to design computers as tools for any purpose, tools that could enable people to think and create in new ways. Believing in the personal computer as a universal tool that can augment abilities and enable people allows for a design vision of multiple possible uses and multiple possible users, including people with disabilities.

Brightman was now in charge of accessibility issues at Apple Computer and, for a while, the sole employee of OSER. In order to make the kind of waves he wanted to for people with disabilities, Brightman knew that Apple's own computers would need to be made accessible, particularly the Macintosh. Brightman describes the Macintosh as "never designed with accessibility in mind."[105] It was more difficult to plug in third-party devices, for instance, as the case on the Apple II was simple to remove whereas the Macintosh was designed with the intention that its users would not remove the casing. In addition, although the new operating system interface was a radical step forward in usability for most people, its new rigorous standards made it so that adaptive software written for the Apple II could not be ported over. Most of the accessibility problems with the Macintosh came from the deliberate black boxing Apple had built into this computer. It was, literally, a closed system, not designed to be opened up and tinkered with the way Apple's earlier computers had been. In the pursuit of user-friendliness, Apple's goal had been to restrict users for their own good; a computer that was harder to mess around with would also be harder to mess up. However, this feature made it more difficult for users whose needs had not been built into it to modify the computer to accommodate their needs. Apple emphasized the user-friendliness of the Macintosh without realizing that restricting users' options would also restrict who the user could be.

In order for Apple's engineers to understand the scope of the accessibility obstacles they had inadvertently created, they had to personally experience the problem. At a meeting with Apple engineers in 1985, Brightman demonstrated the difficulties the Macintosh created for computer users with disabilities. Brightman challenged the engineers to operate a Macintosh and type a memo, using only a pencil held in their teeth, in order to simulate the experiences of a user with severe motor disabilities. The engineers became frustrated by the difficulties in turning on the computer (the power switch was located in the back), loading a disk into the drive, and using a word processor that required multiple keys to be held down at the same time for certain commands. One example of how easy some of these problems were to fix was the beeping noise the computer made in response to user errors. Deaf and hard-of-hearing users had no way of knowing the computer was beeping at them. Once made aware of the problem, Macintosh engineers quickly added a visual flash as an error response when the computer's sound was turned down all the way, thus having error options available via sound or sight for users with different needs.[106]

While not all accessibility features are so easy or even possible to create, many fixes for the Macintosh were this straightforward; all that was needed was an understanding that there was a problem. According to Brightman, "Sixty-three features that afternoon in about three hours were identified; that while they were conveniences to most users, were actually inconveniences or obstacles to different users with disabilities. . . . The lesson there was that most of the accessibility problems were easy to solve; they were not complicated issues. What was hard was knowing that there was a problem, and being reminded that not everyone uses a computer the way you do."[107] OSER's goal of building in accessibility features fit with Apple's values of creating user-friendly computers for individuals.

In addition to the small changes to the Macintosh that made it far more useful for people with disabilities, Apple also advertised the idea that computers can enable people and make the world more accessible to them. In the fall of 1986, it released a public relations video, called "Access," presenting its views on accessible technology and the possibilities for computer users with disabilities.[108] Filled with examples of adaptive technology that allowed computer users with disabilities to communicate, play games, and learn, the video concludes with a segment in which a cartoon wheelchair (similar to the graphic used to represent disabled parking) travels across a street. Upon reaching the opposite sidewalk, the wheelchair bumps into the curb and is unable to continue, as the sidewalk lacks a curb cut. A cartoon construction worker then builds a curb cut, which the wheelchair is able to use to travel up and onto the sidewalk. Following the wheelchair are a number of actual people (not cartoons) using a curb cut to fulfill a variety of needs (a parent pushing a stroller, someone using a dolly to transport heavy goods, a child riding a skateboard, a bicyclist, a shopper pushing a shopping cart, people using roller skates, and even a unicycle). This curb cut that meets the needs of many different people is then compared to a personal computer. In the video, the cartoon wheelchair is shown encountering a staircase made of bricks, which it is unable to ascend. The staircase transforms into an Apple computer. The computer keyboard is shaped like a ramp that the wheelchair is able to travel up and then into the computer screen. Apple thus shows that the personal computer, like a curb cut, can be used by all kinds of people to accomplish all kinds of tasks.

Like the DCCG, Apple claimed the computer offered independence and the ability for people with disabilities to fulfill their desires. Apple also argued that accessible technology benefits everyone; when something is accessible to people with disabilities, it is more usable to all people. By designing a technology with people in mind who will need to use it in varied ways, usability as a whole is increased. These technologies then allow for uses that are not strictly necessary but are beneficial in other ways, such as providing convenience. For instance, by developing computerized conferencing technology with people with disabilities in mind, Murray Turoff and his fellow researchers designed a communication system that could both accommodate disabilities and be useful to everyone.

The separate work done at Apple Computer and the DCCG converged in 1985. That fall, Jackie Brand and Alan Brightman met at a Closing the Gap conference focused on technology in special education. The conference drew many technology developers who showcased products still in development. Brightman gave the keynote speech, offering advice for special education teachers introducing computers into their classrooms. One of his main recommendations was that teachers and parents should talk about what they needed, and the computer industry should listen.[109] After his talk, Brand felt that Brightman would be interested in the work they were doing at the DCCG.[110] Brand and Brightman found themselves to have similar ideas regarding accessible computer technology. Shortly after the conference, Brightman visited the DCCG, and Jackie and Steve Brand visited Apple World. The DCCG and OSER came together at a meeting in February 1986 at Apple, where the DCCG and other disability experts presented to Apple's Education Sales Representatives and engineers on the technological needs of people with disabilities. Fifteen guests and fifteen Apple employees met to discuss how to improve the company's ability to meet those needs.[111] Apple paid the DCCG $500 for giving this presentation and subsequently donated a Macintosh 512 computer to the group.[112] After this meeting, Brightman described the DCCG: "The only thing wrong with them is there aren't more of them around the country. My only fear is that they're going to run out of energy. I think of them as some of the best professional colleagues that I have."[113] This initial collaboration continued, to Apple and the DCCG's mutual benefit. Apple debuted their "Access" video at a DCCG Open House a year later. Around this time, Apple and the DCCG began to collaborate in earnest, as Jackie Brand joined with Alan Brightman to create the National Special Education Alliance.

Chapter Three

Corporate Philanthropy and the National Special Education Alliance

Large, mass-market personal computer companies, such as Apple and IBM, began to build accessibility features into their computers in the mid-1980s, as disability advocates pushed for such features to be a part of computer hardware and software before standardization had built in barriers to accessibility. With the technology now coming into being, advocates within Apple Computer's Office of Special Education and Rehabilitation (OSER) shifted their focus to spreading the word about the possibilities and the realities of personal computer technology for people with disabilities. The technology was being developed; now, it needed to reach users on a national scale. Apple's strategy to promote it involved working with locally based disability and technology advocacy groups, starting with the Disabled Children's Computer Group (DCCG); together, they created the National Special Education Alliance (NSEA)—a national organization that could connect local groups through a network of shared expertise.

Apple's interests in founding the NSEA were based more in corporate philanthropy than in capturing people with disabilities as a consumer market. Apple was not unique in this way; at this time, focusing attention on people with disabilities was mostly a charitable gesture for computer companies. Companies did not conceive of people with disabilities as a potentially profitable market, but as they realized that some people needed accessible technologies to use personal computers and that increased usability could benefit everyone, companies dedicated resources to accessibility. General increased usability made computers more user-friendly—broadening the user base to include people who were unsure of the technology. Employees within computer companies, some disabled themselves, believed that personal computers could specifically help people with disabilities, but in order to do so, consumers and developers needed to be made aware of what was possible. However, the personal computer revolution created an environment where new, better technologies were continually being developed and released to consumers at a pace that made knowledge transmission difficult. Furthermore, this difficulty was compounded by the inherent complexity of these technologies and by the cultural perception of computers as high technology that required certain kinds of intelligence or abilities to operate. Organizations like the NSEA and DCCG sought to combat this perception by communicating knowledge to consumers and teaching them skills. Like Apple, IBM, while developing their own accessible technologies, also worked to get computer technologies into the hands of users through philanthropic programs meant to communicate knowledge about what was available to people with disabilities.

In this chapter, I begin my account of corporate accessibility-related philanthropy and the formation of a national network of disability and technology groups with the creation of the NSEA. Next, I examine the limits of Apple's corporate philanthropy, looking at how the NSEA separated from Apple and became an independent, nonprofit organization called the Alliance for Technology Access (ATA).[1] I also discuss the activities of the DCCG during the late 1980s and early 1990s as it continued its local, on-the-ground activism in the Berkeley area after Jackie and Steve Brand moved on. I conclude by examining the internal accessibility work done by IBM during the late 1980s, which consisted of efforts to distribute computer technology to users and to develop accessible technologies for employees and consumers.

Creation of the NSEA

In the spring of 1986, Jackie Brand and Alan Brightman began to discuss the idea of a national organization that could connect local disability and technology groups such as the DCCG. Brand felt that work being done around the country by local organizations was being held back by their lack of connection with each other.

> One of the things that we saw early on was that there were little efforts taking place all over the world, probably—but to our knowledge, all over the continent—where somebody with a disability had a need and they or a family member or a friend were trying to address the need and create a solution. Those solutions

were kind of mom-and-pop type solutions and they never became part of the mainstream, so other people didn't have access to those solutions. The field was only going to grow and be effective if we could connect people who were working together in the field.[2]

These people and groups looking to share solutions needed a social technology—a network to share and disseminate information and expertise. In order to create ties between organizations and to share resources and solutions, local groups like the DCCG needed to be discovered and brought together, and someone needed to provide funding for an umbrella organization. In the late 1980s, Apple Computer offered the necessary resources to bring this idea to reality.

During the winter of 1986–87, Jackie Brand reduced her hours as the executive director of the DCCG to half time, in order to devote more of her time to working with Alan Brightman at Apple toward the creation of a national disability and computer technology organization.[3] Their project became the National Special Education Alliance. Shortly after its creation, Brightman described their goals for the project: "We established the NSEA to link these groups together for mutual benefit; to help individuals discover working partners; to ensure the timely sharing of information; and, ultimately, to serve the computer-related needs of disabled children and adults within their communities by providing them with information and resources that they may not otherwise have access to."[4] The exchange of knowledge and expertise across the country, while serving local needs, was the core value behind the NSEA. It intended to remedy the difficulty of finding information on computer solutions for people with disabilities, an obstacle that prevented more people from accessing technology that could benefit them.[5] People involved with the NSEA believed in the importance of spreading information about new accessible computer technologies for three reasons:

> First, the development of new technology solutions is occurring so rapidly that parents and professionals alike find it increasingly difficult to keep up with the new possibilities. . . .
>
> Second, these efforts can help schools and parents to invest scarce resources more wisely—for example, by avoiding the purchase of outmoded technology, inappropriate software, or a device of which a competitor has just produced a cheaper and/or better version. . . .
>
> Finally, widening local public awareness of what is now possible technologically helps speed up the process of service delivery. . . .
>
> In other words, public information leads to public action.[6]

The acceleration of personal computer development made it difficult to keep up with the latest technologies, especially for people and groups who lacked extra resources to devote to staying on the cutting edge of innovation. At the same time, consumers became increasingly aware of the potential for personal computer technology to benefit their lives—creating a market for new accessible technologies. What was needed was a bridge between developers and consumers to provide expertise and advice while communicating the needs and desires of users back to developers.

Utilizing the same strategy behind the DCCG, but on a national scale, the NSEA founders believed that the best way to solve the problem of the transfer of knowledge was via partnerships between consumers, disability professionals, and the computer industry: "The key idea is the notion of collaboration between parents and professionals . . . The power of this idea, especially as it relates to populations with special needs, can hardly be overemphasized. [Donna Dutton, director of one of the NSEA centers] also points out that parents and professionals do not easily divide between "us" and "them" on the basis of technology sophistication. Some parents are considerably more sophisticated than professionals on that score, and vice versa."[7] The acknowledgment that anyone could become an expert in computer technology—parents, teachers, even children—kept the NSEA from being modeled on a one-way street, with professionals sharing their knowledge with consumers but providing no way for them to speak back. The NSEA argued that such an environment, where professionals and consumers were not on equal footing, was already causing problems with the introduction of computer technology in special education: "As many special education experts have pointed out, one main impediment to the delivery of educational services to students with disabilities is the way in which parents and professionals relate—poor communication, no communication, feelings by the parents that they are being talked down to or ignored, feelings by professionals that they are being inappropriately challenged or criticized by parents, and so on."[8] The NSEA founders believed that a partnership in which both users and technology vendors played roles would solve some of these communication problems. Brand, in particular, felt that a national alliance, composed of partnerships between consumers and companies, could work to help individuals locally: "NSEA demonstrates how a community/industry partnership implemented on a national level can have dramatic impact on the most local level—changing the world one person at a time."[9] Through a partnership, users and developers would have a place to speak to each other, with the NSEA facilitating the conversation in the middle.

In order for Brand and Brightman's vision of an industry/consumer partnership to become a reality, Apple Computer had to donate time and money to the creation of the NSEA. Apple founded the NSEA at a time when the company had both the interest in and resources to commit to creating such an organization. Jackie Brand described Apple as having a "real commitment to the issues—and not seeing the alliance as some kind of marketing strategy."[10] Apple's dedication to people with disabilities was in line with its stated corporate value of focusing on individuals and improving lives through computer technology. An employee task force memo from September 1981 codified the values the company held as goals.

> We build products we believe in.
> We are here to make a positive difference in society, as well as make a profit.
> Each person is important; each has the opportunity and the obligation to make a difference.[11]

Apple was founded on the belief that technology can change the world for the better and should be shared, and its values were reflected in the philanthropic work that the company did. As it grew into one of the major personal computer manufacturers, however, Apple's founding countercultural and hobbyist computing values struggled against its concern with the bottom line, a conflict that played out in the area of accessibility.

These clashing corporate goals—maximizing profit versus a utopian ideal of shared technology improving all lives—can be traced back to Apple's founders. The different personalities of Steve Jobs and Stephen Wozniak were reflected in Apple's corporate culture, its technological developments, and the tensions within the company. These differences can be seen in the very design of Apple's computers and in its relationship with accessibility. Both Jobs and Wozniak valued elegance and minimalism in computer design. For Jobs, this approach to design was a matter of aesthetics and creating user-friendliness by limiting choices for users. For Wozniak, though, it was about efficiency in using resources so that users would have the option to do whatever they could imagine with the computer. One of his original visions for personal computers was that they would be flexible: "I had the idea that there would be a lot of things people would want in the future, and no way did we want to limit people."[12] The Apple II enacted this goal of openness, but the company moved away from it with increasingly black-boxed designs, starting with the Macintosh. Black-boxing their computers in the name of user-friendliness was the result of the same paternalistic attitude evident in Apple's early dedication to accessibility; by restricting users' options, the company was attempting to constrain what users could do.

These different understandings of Apple's core values led to clashes in technology development and between employees. For example, Jef Raskin, the initial head of the Macintosh project, wanted the new machine to be affordable, while also being user-friendly and functional.[13] Once Raskin was off the project, John Sculley set the price of the Macintosh much higher than competitor's computers, in order to gain short-term profits at the eventual cost of a greater market share.[14] Such clashes would lead uncompromisingly idealistic employees, such as Raskin, to eventually leave the company out of frustration when decisions were made above them.[15]

While the company was unable to always stick to its more utopian ideals, there were occasional moments when Apple lived up to its core values. For example, Raskin's intention for the Macintosh to empower people and Sculley's instruction to Brightman that if he was going to fail with the Office of Special Education, then he should fail big, both embody Apple's founding value—that the potential of computer technology is that it can change people's lives for the better. These values again surfaced with Apple's support for the NSEA. During testimony before Congress a few years later regarding passage of the Technology-Related Assistance for Individuals with Disabilities Act, Apple stated, "The corporate commitment by Apple Computer, Inc. toward the advancement of technology for use by individuals with disabilities is powerful, enduring and passionate."[16] Apple was one of the first major computer companies to have an internal group dedicated to people with disabilities and accessibility, in the form of its Office of Special Education and Rehabilitation, which took the more general name of Worldwide Disability Solutions in the late 1980s. Supporting people with disabilities fell in line with Apple's spoken mission of making computers for individuals; by taking into account as many uses of the technology as possible, the needs of as many individual users could be accommodated. Even with this focus on individuals, however, Apple still treated people with disabilities differently from other users in one key way: they were not seen as a potential consumer market that could profit the company. Instead, Apple targeted computer users with disabilities out of a notion of doing the right thing. Apple's paternalistic philanthropy toward people with disabilities allowed the company to live up to its founding values in a bounded way, controlled and circumscribed by its view of who its users were.

Apple's dedication to users with disabilities, even if the company did not quite see them as customers, had a solid, long-lasting role within the company, though it would peter out in the late 1990s. Apple continued to supply resources to the resource centers after the initial creation of the NSEA, as advertised in a promotional brochure from 1988: "To these charter-member centers—and to all new resource centers that join the Alliance—Apple offers ongoing assistance in the form of computer equipment, resource materials, technical assistance, and moral support."[17] Apple provided such assistance to the NSEA for another few years. During the time that the NSEA was an Apple project, John Sculley was CEO of the company. For much of the decade under his leadership, Apple experienced success and stability, which allowed the company to dedicate resources to its philanthropic concerns such as aiding people with disabilities.

NSEA Charter Members

The goals of the NSEA could only be realized once likeminded organizations were located to combine under the Alliance's umbrella. While Brand and Brightman were working out the big picture of what the NSEA would be, Brand also began to travel around the United States with an Apple OSER employee, Robin Coles, to find local organizations interested in joining the Alliance.[18] "So along with Robin Cole[s], who was at Apple at the time, I traveled for about six months and we met groups and families and people who were getting things started, who wanted to get things started, who were looking for hope, who were excited about the possibilities, and through that whole process of beginning to connect up a national community, started to really build the national organization."[19] Brand and Coles represented the kind of partnership between consumers/activists and developers/companies that the NSEA sought to create, a partnership they then used to locate and gather organizations with the same goals and values.

The NSEA explained the ideals behind their partnership with those local groups chosen to join and insisted on certain criteria for future member centers: "Acceptance in the Alliance requires that each resource center be a genuinely collaborative venture among people with disabilities, their parents and friends, and professionals in the fields that serve them (education, rehabilitation, and so on). Each resource center is led as much by parents of disabled individuals and the individuals themselves as by professionals."[20] In addition to the requirement of partnership, the local centers also had to be able to operate successfully and independently within their communities. "Their proposals for charter membership in the Alliance were accepted only after a series of on-site visits convinced representatives of Apple and DCCG that they were (1) capable of providing technology leadership in their areas, (2) more broadly based than an exclusively professional agency, (3) committed to the goals of the Alliance, (4) able to generate local support, and

(5) enthusiastic about the potential of microcomputers to change the lives of people with disabilities."[21] The centers needed to be viable, able to accomplish necessary goals, and willing to operate in line with NSEA ideals.

As a benchmark of what the NSEA was looking for in a local disability and technology resource center, Apple used the DCCG as its model: "Apple defined its role as a catalytic agent, working with DCCG to try to identify places around the country where seeds of a DCCG-like operation might be planted." Apple initially chose ten resource centers across the country to join the DCCG in the Alliance. These centers were in ten different states: California (DCCG and Computer Access Center), Colorado (Children's Hospital Resource Center), Florida (Computer Center for Independent Technology and Education), Illinois (Technical Aids and Assistance for the Disabled Center), Kansas (Technology Resources for Special People), Kentucky (Disabled Citizens Computer Center), Massachusetts (Massachusetts Special Technology Access Center), Minnesota (Parent Advocacy Coalition for Educational Rights Center, Inc.), Nevada (Nevada Computer and Technology Center for the Disabled), and Ohio (Communication Assistance Resource Service).[22]

From the corporate side of the NSEA partnership, Apple continued its philanthropic dedication to people with disabilities through donations of various types to the charter member organizations. The company gave Apple IIs and Macintoshes to each center, set up and maintained computer-based telecommunications between the organizations using AppleLink, and paid for organization members to travel to national conferences. To send a message to communities about what Apple wanted the NSEA to represent, the company also provided each charter center with a coffee urn, as a way, Brightman said, "to get the centers' directors thinking about how they can more effectively reach out into the community and bring teachers and parents into the centers."[23] In a more recent interview, he recalled donating the coffee urns because they "said to people this is a place for people."[24] The NSEA intended the environment at the local centers to be one of community and involvement, echoing the value of partnership at the core of the Alliance. This type of humanizing gesture acted to normalize the computer technology environment of the resource centers by providing something as everyday as coffee for visitors, much as the name "Apple" helped to make the company seem friendly and nonthreatening. These were steps toward making people think of the complex technology of the computer as an everyday technology.

There were four components to the NSEA's strategy for communication and networking between the centers; the first two dealt with computer networks. "First, the centers have been electronically linked to one another via AppleLink, Apple's information and communications network. . . . Second, each center has been linked to major national databases and bulletin boards via both AppleLink and SpecialNet, the largest telecommunications service in the country devoted specifically to special education and rehabilitation."[25] Providing the centers with AppleLink and access to SpecialNet took advantage of early networking capabilities to connect the centers to each other and to the outside world.

AppleLink provided e-mail and instant-messaging capabilities between computers located at the different NSEA centers. AppleLink had debuted only a few years earlier, in 1985, as the first online service with user-friendly windows and icons instead of a command-line interface.[26] Run as a joint project between Apple and General Electric (GE), it was intended for employee use only and was not available to the general public. Apple offered a personal edition of AppleLink for the public in 1988; it would, through a partnership with a company called Quantum, go on to become America Online.[27] In an interview, Brand told a story demonstrating the effect AppleLink had on what the centers could accomplish through distributed communication.

> We would get emails like, "A family just came in and the kid is trying to do math and he's really interested in race cars, and he's not able to really handle the computer, and help! What ideas do you have?" And within twenty-four hours from all over the country, people would throw out everything they knew, everything they had come across from, gosh, "There's somebody in a small town in Kansas who's just developed this great little car-racing program that has math built into it," to, you know, "I just met with somebody and here's a way I used a new keyboard—it was very successful."[28]

This kind of exchange of information through networked computers also took place on a larger scale. SpecialNet offered around one hundred different electronic bulletin boards that people all over the country could connect to via a modem. It brought together special education offices in every state, two thousand school districts, hundreds of colleges, and various other programs.[29] The use of this early networking technology showed that disability activists were at the cutting edge of computer technology, embracing it as it developed to find the best technological solutions for people with disabilities. Networking technologies like this demonstrated the new kinds of abilities that personal computer technology made possible.

In addition to implementing electronic networking at the centers, Apple also set up ways for NSEA members to interact directly with vendors and to meet together in person. These were the last two components in their strategy:

> Third, Apple has created a category of membership in the Alliance called "charter member organizations"—professional organizations, technology vendors, publishers, research and development centers, rehabilitation centers, and the like. These organizations have agreed to keep the resource centers informed about new and forthcoming products and ideas, and to cooperate with individual centers in identifying appropriate solutions, training local people in the use of technology devices, and making experimental materials and ideas available to the centers for field testing.
>
> Finally, recognizing the importance of initial face-to-face contacts among individuals working in the centers, Apple has initiated a series of national meetings designed both to provide training and orientation to the members, and to give them the all-important opportunity to get to know each other as well as many of the participating vendors and professional organizations.[30]

Face-to-face meetings mainly took place at conferences related to disabilities and technology, giving resource center representatives opportunities to meet and to connect with others working on related projects external to the NSEA. Brand described the important role this electronic and face-to-face communication network played in what the NSEA was capable of accomplishing: "We basically gathered information that was otherwise unavailable. It wasn't documented anywhere. It was direct experience and feedback and sharing of resources that all of a sudden gave us a sense that there was now a national focus on issues we had been struggling with, one by one, in our own little programs, in our own homes, in our own little centers."[31] These different aspects of the communication network Apple established, which consisted of electronic-based and face-to-face communication with other centers and technology vendors, allowed the NSEA to create a larger set of information than any individual center could achieve alone.

Spread across the country, this shared pool of expertise made the NSEA more than just a collection of local centers. A distributed information network allowed the NSEA centers to stay small and locally focused while finding solutions to problems beyond their limited expertise. With its access to the AppleLink service, the NSEA functioned as a computer network, with each center able to contact the others online. Before there was widespread public access to the internet, this network allowed the NSEA centers to operate and communicate with each other in a way that computer networking technologies would, in a few years, allow everyone to access information at a distance. This computer network also allowed direct communication with other people through their computers. Murray Turoff's late 1970s vision of future communication through computer conferencing began to be enacted here, and AppleLink was a far more user-friendly system than the technology Turoff had originally envisioned. As Turoff had hoped, however, computer networking technology was being used to benefit the lives of people with disabilities. The development of these cutting-edge technologies collided with disability issues as personal computers granted new kinds of abilities to users. People with disabilities were the paradigmatic computer users, embodying the need for the new capabilities the technology could provide and demonstrating how the personal computer could augment all human abilities. Accessible computer technologies were the first step in developing machines that were usable for everyone, an important step along the road to the fulfillment of the computer as the universal tool.

Benefits for Resource Centers and Technology Vendors

In the spring of 1987, the NSEA project officially launched, with eleven local resource centers and fifty-three professional organizations and third-party vendors of computer technology.[32] Resource centers and vendors saw immediate benefits from being a part of the Alliance, and these benefits kept them working toward the organization's success. From the perspective of the local centers, being connected directly with Apple had advantages beyond the material donations the company made. The resource centers found that Apple's name on the project drew greater attention from other organizations. Carol Cohen, of Computer CITE in Florida, described the increased generosity her local community showed after CITE joined the NSEA: "If you mention that you have support from Apple, people want to help you. Apple is like a magic word."[33] The centers did not only have Apple to rely on, however; they also had access to each other through the communication system Apple had set up. The NSEA found that the diversity of the various centers allowed technical expertise to be distributed across the various organizations: "One of the strengths of the Alliance as a whole is the difference in the initial strengths of the members, since this enabled members to get help from one another. For example, one center has special expertise and long experience in developing technology solutions for people with visual impairments, another is especially knowledgeable about communication aids and devices, and a third has worked intensively to develop technology solutions for young children with severe disabilities."[34] Once the NSEA was established, the centers could immediately pool and rely on the expertise of each other to solve problems and aid the disabled population more effectively than they could when they were isolated.

Although Apple was responsible for creating the NSEA and donating essential resources to each local center, the NSEA was not an Apple-exclusive organization. From the beginning, even though the NSEA was deeply connected to Apple, it was not a direct branch of the company. In numerous NSEA promotional materials and published articles, Apple described the hands-off approach they took with the centers and the lack of any requirements that Apple products be promoted or used: "The equipment needed by an individual is the equipment that needs to be used, whether it's Apple, Atari, or whatever. The resource centers are not Apple centers; and they are not run by Apple Computer."[35] The NSEA enforced Apple's public promise that "the Alliance, then, is not a division of Apple" in the way the organization structured its decision making.[36] The Alliance Planning Team (the initial planning group of the NSEA for its first few years) was required to always include parents, consumers, professionals, and vendors among its representatives. Brightman further decided that Apple would only have one voting member in the group.[37] This distancing of Apple from overt control of the NSEA was a manifestation of the company's philanthropic, though paternalistic, attitude toward the Alliance and people with disabilities. Apple did not attempt to use the NSEA as a way to attract people with disabilities as a profitable consumer market but instead seems to have viewed them, in this context, only as a group who could benefit from Apple's goodwill.

Apple and the third-party vendors who joined found that the NSEA brought them benefits as well. Apple and the other companies found their reputation improved among people with disabilities and disability professionals. By generating increased use of computer technology among people with disabilities, the NSEA helped organize people with disabilities as consumers to whom computer technology companies would eventually try to appeal.[38] The NSEA did not only disseminate information to potential users but also communicated back to developers. The idea of partnership at the heart of the NSEA created an opportunity for technology companies to gain direct feedback from consumers about their needs: "these grassroots organizations, now connected in a meaningful way, suddenly have a greater potential value to the vendors than they had as individual organizations. They offer a source of unified consumer feedback, provide ideas for new products and programs, act as 'showcase sites' for products, and (increasingly as the Alliance grows) provide a means to reduce the lag time between the arrival of a new product on the market and awareness of its existence among special-needs purchasers."[39] The exchange of knowledge between members of the Alliance also influenced technology development, as vendors had a new way to communicate with users of their products. Small, third-party vendors found it easier to communicate with consumers through the resource centers as a part of the NSEA, and they also discovered that they had access to greater attention from Apple: "As Dr. Mary Wilson, president of Laureate Systems, describes it, the connection of the centers with Apple Computer encourages vendors to be cooperative. 'The centers have much higher visibility because of the Apple connection,' she reports. 'They also give us a way to indicate to Apple that we want to work with them. As a relatively small vendor, we might otherwise not get as much attention from Apple technicians and programmers.'"[40] Another motivation behind the NSEA was quicker development of products. The communication system the NSEA established placed it in a position as a go-between for users and developers, and the system enabled members of both groups to interact in such a way that both were on more equal footing with each other.

The Alliance Becomes Independent

Once the NSEA was up and running, its leaders quickly looked to make it independent from Apple. The NSEA needed to separate itself in order to best connect users with computer technology that could benefit them; the organization needed to be free of corporate ties that might limit what it would be allowed to do. The Alliance could not effectively function as a bridge between users and developers if it fell under the auspices of one of the major computer companies. So, during the two years after it was formed, the NSEA worked to become an independent, nonprofit organization. In the spring of 1988, Jackie Brand formally left her position as executive director of the DCCG to work full time at Apple Computer, as executive director of the NSEA project.[41] That fall, NSEA employees began to meet with Apple's lawyers to look into legally separating the Alliance from Apple.[42] Though Apple showed no signs yet of wanting to stop supporting the NSEA, the company was willing to let the project go. As a part of the changing relationship between Apple and the NSEA, the Alliance Planning Team instructed NSEA centers to prevent possible conflicts of interest by ending the sale of Apple computers directly through the centers.[43] On February 28, 1989, the group became its own corporation, operating under the name NSEA Foundation, and filed for nonprofit tax status. Brand described the relationship between the NSEA and Apple in the application for the group's nonprofit status: "Apple's experience with this project has led it to conclude that a national foundation, which is not controlled by Apple, may benefit people with disabilities by making it more likely that corporations other than Apple will support it. Therefore, Apple has encouraged the formation of the NSEA Foundation as an independent entity. Apple will provide start-up support to the Foundation. However, Apple will reduce its role as the Foundation develops more broadly based support."[44] Though Apple's oversight of the NSEA was diminishing, Brand was able to use funding from Apple to hire the first three staff members of the nonprofit: Donna Dutton, head of the NSEA member Computer Access Center in Santa Monica, California; Harvey Pressman, author of most NSEA publications; and Robert Glass, cofounder of the Disabled Citizens Computer Center in Louisville, Kentucky.[45]

By the fall of 1989, the NSEA was a completely independent nonprofit organization, and the Apple Computer logo had been removed from its materials. All NSEA resource centers were required to have nonprofit status. In order to more accurately reflect the goals of the NSEA, which was concerned with more than just special education, the organization changed its name to the Alliance for Technology Access (ATA).[46] The ATA set up a new national office in Albany, California, in order to further establish its independence from Apple and the DCCG, which had previously provided space for the NSEA.

Jackie Brand viewed the ATA's separation from Apple as an inevitable step toward the Alliance becoming the kind of organization it wanted to be: "Our missions were distinct: the bottom line for Apple was to make money, the bottom line for the Alliance for Technology Access was to provide technology and information about technology to people with disabilities. I always worried about the day when it would become not so interesting to Apple anymore, and we were always trying to prepare for that eventuality."[47] Brand was correct that their discrete goals would eventually cause Apple to stop supporting the ATA. In the late 1980s, people with disabilities were still a philanthropic concern for Apple, rather than a strong market segment, and the company began cutting back on the resources it donated to the ATA and its member centers. Glass described the dwindling support: "From this time onward, poor computer sales industry wide and repeated challenges internally from higher management authorities at Apple Computer over the value and wisdom of investing dwindling marketing budgets in the field of disability cause Apple's corporate contribution to the Alliance to begin diminishing."[48] As a consequence of this divestment, in early 1990, OSER cut its support of AppleLink between the Alliance centers because it lacked the budget to continue funding the project, and the ATA took over funding the AppleLink connection.[49]

As the ATA grew and began to support itself through contributions from sources other than Apple, it turned to larger projects, moving beyond local work and connecting discrete groups to national concerns. The network the ATA constructed was capable of expanding to include different ways of reaching computer users with disabilities beyond direct one-on-one contact between a client and a member center. At the top level of the organization, work done by the ATA involved more than just maintaining a communication network between resource centers. The Alliance's dedication to using computer technology to break down barriers for people with disabilities extended to projects involving education, training, funding, and technical support for the centers and their clients. In the Alliance's application for nonprofit status, Brand listed many of the types of programs the organization intended to pursue.

- Educating and training people with disabilities, their families, and service providers in the use of computer-related technology to remedy specific problems caused by various disabling conditions;
- Developing educational materials so that other exempt organizations can offer similar programs;
- Funding the training of service providers, people with disabilities, and family members in the use of computer-related technology;
- Making grants to exempt organizations to develop service for people with disabilities;
- Making grants to individuals for further study and training in the use of computer-related technology by people with disabilities;
- Developing, or funding the development of, technological solutions to problems encountered by people with disabilities; and
- Providing technical assistance to other exempt organizations in such areas as outreach to minority people with disabilities, preschool and K–12 programs, organizational development, and program development.[50]

The ATA carried out many of these activities as it grew. Within two years of its founding, the ATA included thirty-eight local technology resource centers in twenty-eight states; by 1992, there were forty-five centers in thirty-four states.[51] Since the early 1990s, the ATA has included between forty and forty-five centers under its umbrella. The Alliance has been unable to grow much larger, a limitation that appears to be based in the ATA's network structure, which is composed of many small, local organizations. These organizations frequently rely on the energy and dedication of the individual people running them and are susceptible to failure when these people lose interest in maintaining them (a situation I examine further in chapter 4). However, despite occasional problems with centers failing and dropping out, the ATA has managed, as a whole, to stay stable and maintain its mission since its founding.

The DCCG after the Brands, 1987–1991

While the ATA matured into a national presence in disability and technology advocacy, the DCCG continued to provide services for individuals with disabilities at a local level. The DCCG acted as a counterpoint to the ATA; whereas the ATA worked on large projects, connecting groups across the country together, the DCCG continued to function locally. It was the place where users came in personally for help, and, as such, it makes a good case study for examining the immediacy of interactions between computer users with disabilities and accessible computer technologies. Unlike other small, local organizations that crumbled when their founders move on, the DCCG remained strong after the Brands left, and it continued as a successful local organization, enacting the ideals of the ATA. Kate Sefton and Linda De Lucchi took over for Jackie and Steve Brand, in 1988, acting as executive director and president, respectively. De Lucchi was a former employee of the Lawrence Hall of Science, where the DCCG was founded, and one of the DCCG's first members. Sefton was a professional developmental therapist who had immediate family members with disabilities and had seen firsthand how computer technology was able to impact their lives for the better.[52] Sefton's perspective on disseminating knowledge to users emphasized the necessity of keeping pace with changes to the technology. In order to stay on this cutting edge, the DCCG needed space where people could interact directly with the technology.

In 1988, the DCCG moved its resource center into a much larger space on Rose Street in Berkeley, where it would remain until 1992. This increased space allowed dedicated room for administrative uses, computer and other technology storage, and space for large events. By 1989, the DCCG had divided its programs into five categories: the Interactive Computer Resource Center, Technical Problem Solving, Training, Loan Programs, and Communications.[53] These various programs demonstrate the strategies the DCCG attempted in order to bring people with disabilities together with personal computer technology; in the same way that accessible technologies needed to meet the needs of different individuals, the DCCG targeted its communication with users based upon what they most needed or wanted out of the technology.

The Interactive Computer Resource Center regularly held meetings, open houses, and workshops for large groups of people to come in and learn about what the DCCG offered and what technology was available. For example, in December

1991, the DCCG held a workshop on Toy Adapting and Switch Making. The center also performed outreach to underserved populations in the area. Technical Problem Solving programs focused more on one-on-one or small group interactions, and they targeted families, professionals, vendors, and special interest groups within the DCCG. Such groups, which appear to have met monthly, included a Visually Impaired Interest Group, an Interest Group on Technology for Persons with Severe Disabilities, a Unicorn Users Group, a Dialog with Developers group, and a HyperCard Interest Group. By 1991, an Augmentative Communication User Group and a Computer Maniacs group also formed. The former, in particular, held frequent meetings (up to two or three times per month) and had separate subgroups for adults and children. These groups included people beyond DCCG staff and interested users; Arjan Khalsa from Unicorn Engineering and Mike Palin from Words Plus came in 1989 to work with Special Interest Groups dedicated to their respective technologies.[54] In these groups, users and developers could come together, with the DCCG acting as a channel for communication. Users groups also put an emphasis on tinkering, with users solving problems for themselves and sharing solutions with each other.

Training activities included setting up conference presentations and connecting the DCCG with school district staff who needed help with technology. Kids on Keyboards, begun in 1988, was a popular and oft-mentioned DCCG program that attempted to improve the local community's technical expertise: "Youngsters meet in an informal atmosphere, learning computer skills while making friends and sharing a joke. Kids on Keyboards is only possible because of the dozens of volunteers that work, learn, and laugh right along with the children."[55] The DCCG also had plans for future training programs related to leading small seminars and video production. By late 1991, the group began to hold small seminars on topics such as Unicorn keyboards, Macintosh computers, and technology for beginners.

The DCCG ran loan programs through which people could borrow materials such as software, adaptive devices, printed resources, and instructional videotapes. In addition, the organization had modems they would loan out so that people could access a free electronic bulletin board run by the group.[56] These programs allowed the DCCG to account for the different kinds of embodied uses their clients experienced and provided a means for users to test various technologies to discover what fit them best. In 1987, Pacific Bell donated a larger space for the DCCG Resource Center; here, the organization was able to set up one of their most popular loan services, the Toy Lending Library. Run by Alice Wershing, a special education teacher and member of the DCCG steering committee, the library maintained a collection of toys and games that included advanced computer technology and simple mechanical or traditional toys.[57] Wershing worked with children (from infants to nearly teenagers) and their families to test and play with the toys so that they could figure out which ones the children most enjoyed and which ones could be made to accommodate their individual disabilities.[58]

These various lending programs helped to fulfill Steve Brand's goal that the DCCG should find better ways to connect people with technology that could benefit them. In 1987, an NSEA booklet described his perspective on the importance of being able to try out technology: "One thing that's become important to [Steve Brand] is to increase and improve the lending of hardware and software to members. He wants to be able to say, 'Here, take this home. Try it out. See if it works.' He wants to find ways to make it easier for people who still think the computer is a 'longshot' in helping their children, who don't yet realize all the ways it can brighten their futures, who feel it is 'too hard' to understand all this new technology."[59] By the late 1980s, the DCCG's expanded loan services allowed for people to test out and play with cutting-edge accessible computer technology on their own time, as a means to familiarize themselves with what was out there and to find solutions for their individual embodied uses. This allowed for experimentation with technology in a way that was otherwise unavailable for consumers.

Finally, DCCG communications programs sent technical information and product referrals by phone calls, mailings, and a regular newsletter. By the early 1990s, the DCCG had clearly expanded beyond its early focus to address the wider interests of children and adults with disabilities, and it was providing resources and training for people with all levels of computer expertise and different kinds of disabilities. It stayed small and locally focused, while at the same time taking advantage of the exchange of knowledge and expertise that being a member of the ATA provided. The DCCG provided face-to-face contact with users, working directly with them to find technological solutions to their individual needs. Within the ATA, the DCCG—and other member centers like them—functioned as separate nodes, connected by the communication network the ATA created. That network linked each center with all the others and connected those centers with technology vendors.

IBM Programs and Technologies for People with Disabilities, 1984–1991

Apple Computer was not the only large computer company dedicating resources to accessible computer technology during the late 1980s; IBM replicated some of the same philanthropic efforts in the form of its National Support Center for People with Disabilities and developed its own accessible personal computer technologies for its employees and customers. To make it easier for people with disabilities to obtain technologies, IBM ran the Offering for Persons with Disabilities program through its National Support Center, during the late 1980s and early 1990s. This program operated through local disability service organizations to provide computer technologies (IBM Personal System/2 and related products) at a discount.[60] The discounts offered were between 33 to 50 percent off and were intended to be used for purchases made for rehabilitative purposes. The local service organizations—mostly branches of the Easter Seals or United Cerebral Palsy Association—helped people using the program to select, order, and install IBM products.[61] Like Apple, IBM worked with these external disability advocacy groups to reach people the company believed would benefit from its products.

A 1988 bulletin in the ACM Special Interest Group on Computers and the Physically Handicapped newsletter described the necessary qualifications for individuals with disabilities to acquire computer equipment through the program.[62] Initially, the discounts were only available to residents of ten states through the Easter Seals or United Cerebral Palsy; in 1989 the program expanded to include other independent local organizations.[63] By 1990, there were thirty locations across the country where people could participate in the program, which required them to supply certification of their disability from a physician. The physician also needed to show that the computer equipment being asked for would provide a "therapeutic or rehabilitative benefit" to the user.[64] In terms of the types of disabilities that would qualify someone for the program, IBM kept the requirements broad and inclusive, including people with a "visual, auditory, physical, neurological, learning, or developmental disability; with a communicative disorder, and/or with a similar disability or disorder."[65] Eligible users could purchase one computer system per year for personal use only. As described in the SIGCAPH newsletter, local Easter Seals Service Centers would determine applicants' eligibility, consult about technical equipment for intended users, recommend additional adaptive devices they might need, help place the orders with IBM, and provide computer installation assistance and ongoing support.[66] A few years later, IBM expanded its efforts with external advocacy groups and began working directly with the ATA.

Beyond providing ways for people with disabilities to acquire personal computer technology that might benefit them, IBM also developed its own accessible technologies in-house. The creation of these technologies was different from the philanthropic efforts of Apple. IBM's efforts started from pragmatic concerns to make computers accessible for its own employees, which grew into marketing technologies for the public. As computer companies began to consider people with disabilities as consumers rather than as people in need of charity, IBM started in the late 1980s developing products for its Independence Series. Prior to this collection of software, there were two significant IBM technologies that made the PC more accessible for people with disabilities. Jim Thatcher, a mathematician at IBM's Thomas J. Watson Research Center in New York, developed the Screen Reader, the product from which general screen-reader technology evolved. As with many accessible technology developers and activists, Thatcher became involved in computer accessibility for personal reasons: his thesis advisor and fellow IBM colleague, Jesse Wright, was blind and needed better access to the IBM computers he worked with. In 1984, the two of them began to work on an "audio access system" that could read aloud text displayed on the screen.[67] They named this product PC SAID, after the SAID (Synthetic Audio Interface Driver), a prototype talking terminal developed in 1978 that gave blind users access to the IBM 3277 computer mainframe system.[68] Their work on PC SAID would become the IBM Screen Reader for DOS two years later. Thatcher did not set out to create a commercial product: "I had no idea it would become an IBM product because I was just having fun, making the PC accessible for Jesse."[69] As the first Screen Reader was created to make IBM computers accessible for its own employees, it was not a trademarked product. The Screen Reader would later evolve into the Screen Reader/2, the first screen reader for IBM's graphical user interface operating system, Operating System/2.

The second accessible technology created for IBM personal computers was AccessDOS, a free suite of keyboard accessibility features that ran on DOS and worked with the IBM PC, PS/2, or compatible computers. AccessDOS was developed by the TRACE Center, at the University of Wisconsin–Madison, with funding from IBM and the National Institute on Disability and Rehabilitation Research, and was released in 1991.[70] The suite included many of the accessibility features that were also built into the Macintosh during the mid-1980s, the kinds of flexible user-experience tweaks that later became integrated into operating systems as standard options. AccessDOS allowed alternate ways of using the keyboard, which was particularly useful for people who had trouble with hand coordination or who could only press one key at a time. Features included in the suite were: StickyKeys (turns multi-key commands into single key presses), MouseKeys (controls the cursor with a keypad instead of the mouse), RepeatKeys (adjusts how quickly a key repeats when pressed down), SlowKeys (adjusts how quickly the computer responds to a key press), BounceKeys (prevents the computer from responding to an accidental double-tap of a key), ToggleKeys (provides an audio indication when lock keys, such as caps lock or number lock, are pressed down), SerialKeys (allows adaptive input devices to be recognized through the computer's serial port), ShowSounds (provides a visual display when the computer makes an error beep), and TimeOut (allows AccessDOS features to turn off, so that a computer can be shared with users who do not need the features).[71]

In 1987, the IBM Personal System/2 (PS/2) replaced the PC as IBM's leading personal computer system and was its first with a graphical user interface. A focus on accessibility within the company led to some features becoming integrated into the PS/2 as standard instead of only being available as aftermarket add-ons or through tinkering with the technology. For example, the machine had its power switch on the front of the case instead of the back, making it easier to reach—a feature that people with mobility impairments had been requesting since the first personal computers. The PS/2 keyboard also had nibs on the F and J, a tactile feature that helps people with visual impairments quickly place their fingers on the correct keys and that has become a standard that benefits all computer users.[72] It also came with a computer monitor that could be tilted to allow the screen to be adjusted to any angle—a feature designed for people with cerebral palsy. According to an article in *Think,* the PS/2 had thirty-two accessibility features built into it, most of which benefited all users of the computer.[73] This awareness of how accessibility features made the computer more usable for everyone increased during the 1980s and 1990s, culminating in the tenets of universal design in 1997.

The successor to the IBM Screen Reader, the Screen Reader/2 for the PS/2 personal computer, was developed by IBM's Special Needs System group between 1988 and 1991 and was marketed by its National Support Center for Persons with Disabilities. It was the first product in the Independence Series and marked IBM's turn from creating screen readers mainly for its own employees to developing them as a consumer technology.[74] Thatcher led the development of Screen Reader/2, which was again tested by and initially designed to fit the needs of blind IBM employees.[75] One of the features that emerged from user feedback was Autospeak, which allowed the screen reader to detect changes on the screen, such as error or status messages, and automatically read them to the user.[76] Thatcher communicated with the larger blind public during the development of Screen Reader/2, demonstrating prototypes of the technology at National Federation for the Blind conferences in the early 1990s.[77] Users operated Screen Reader/2 via a separate eighteen-key keypad, so that the keyboard would not be occupied with control of the screen reader (though the user had the option to use the keyboard if they wished). Users also had control over how much text they wanted read at a time: the entire screen, a paragraph, a sentence, individual words, or even one character at a time. IBM designed Screen Reader/2 to work well with different kinds of software and allowed different application profiles to be set up for each program the user wanted to run.[78] It came with already built-in profiles for some common programs. Working with the graphical user interface, it could recognize and translate icons and cursor placement.[79] Screen Reader/2 was also able with its keypad to emulate the function of a mouse by allowing for point-and-click navigation.[80] Although the IBM PS/2 and its operating system, OS/2, would quickly be supplanted in the personal computer market by machines running Windows, IBM's work developing the Screen Reader/2 in communication with the blind community helped to clarify their needs as users and set a bar for other computer developers to match in meeting the needs of the diverse user base.

The second Independence Series product was SpeechViewer, which was developed to be used during therapy for people with speech disabilities. It was the product of nearly a decade of research done by IBM's Paris Scientific Center. As with Screen Reader, IBM tested SpeechViewer with its intended users during the development process, taking it to 150 locations, including hospitals and schools for deaf people around the globe. IBM intended SpeechViewer to support traditional speech therapy by providing the client with computer feedback to different aspects of speech. Its developers described it, in a 1989 conference presentation, as using "gamelike strategies" to encourage the client's progress. There were three types of modules built into SpeechViewer: awareness, skill building, and patterning. Awareness dealt with simple cause-and-effect reactions to sound, loudness, and pitch. Skill building involved the computer providing feedback to nonlanguage voice aspects, such as pitch, "vowel accuracy," and "vowel contrasting." The patterning modules matched "visual representations of speech attributes" by displaying different representations of aspects of speech.[81]

The third Independence Series product was the IBM PhoneCommunicator, developed to allow deaf people to use a computer to engage in telephone conversations. It worked alongside standard Telecommunication Devices for the Deaf to translate conversations into text and display them on the screen. It also allowed communication between deaf people through the computer. At its core, the Phone-Communicator was similar to earlier computer telecommunication technologies for deaf people, such as those surveyed by Peter McWilliams and Frank Bowe in their mid-1980s books on computer technology for people with disabilities.[82] IBM's product did add some new features, however, such as the ability to act as an answering machine and automatically record messages for the user.

The next Independence Series product, THINKable, was similar to Speech-Viewer in that it was developed for medical professionals to use when treating people with disabilities, in this case for people with cognitive disabilities. THINKable provided skill practice exercises to be used during therapy and case management capabilities for medical professionals to manage client details. Its focus was on improving four aspects of memory: Visual Attention, Visual Discrimination, Visual Memory, and Visual Sequential Memory.[83] THINKable used multimedia clips, such as pictures and recorded speech, to provide different sensory stimuli. The case management functions included data collection, analysis tools, and reporting.

The final Independence Series product developed during this time, VoiceType, was released in 1994. It was a voice command system for a personal computer that allowed the user to navigate and control the computer using speech instead of a keyboard and mouse. VoiceType was available for DOS, OS/2, and Windows. In 1996, IBM offered a popular and simplified version of the software, VoiceType Simply Speaking for Windows 95, intended for home and school use.[84] VoiceType was capable of adaptive learning to fit each user's unique speech patterns, and it allowed multiple users to store their profiles in one system. It possessed a vocabulary of more than 22,000 words. VoiceType Dictation, the professional version of the software, came with specialized vocabulary for journalists, doctors, and lawyers.[85] It was available in English, Italian, Spanish, German, and French. As with Screen Reader/2, VoiceType came preset with commands for popular software applications, such as Lotus Notes, Lotus 1-2-3, Microsoft Excel, Word, and Quicken, and included accessible documentation so that users could access help files on their own. Users could perform complicated actions by setting up their own multistep commands using voice macros that could store up to 1,000 keystrokes in a single voice command.[86]

The accessibility work done within IBM during the 1980s and 1990s was, to a large extent, focused on IBM's employees or on training people with disabilities for computer-related careers. The company slowly worked to produce technologies to benefit all people with disabilities, developing products that allowed people to access personal computers and software that was designed to augment therapy. IBM's accessibility work brought users into the development process; employees with disabilities who might use the products being developed and external potential users tested products and supplied feedback on desired features. IBM designed its products so that users could work with them independently, with software documentation and help files made to work in the same manner as the software, whether they were read aloud to the user with Screen Reader or navigable by voice with VoiceType. The company's accessibility efforts, however, were not only internally focused. It had contact with and participated in projects with external disability and technology organizations across the country—connections that would grow in the early 1990s.

Both IBM and Apple dedicated company resources to improving personal computer accessibility for people with disabilities during the 1980s and 1990s. There were, however, a number of significant differences between the two companies that affected how they developed accessible technologies and worked with disability activist organizations. IBM was, of course, a far larger company than Apple. It developed a large variety of computer products, while Apple focused on personal computers. Whereas IBM had existed since the early twentieth century and developed some of the earliest computer technology, Apple was founded with the invention of the personal computer in the late 1970s. IBM's development of personal computers and work with people with disabilities were a part of the company's long history, and many of the company's computer accessibility features were developed internally in order to make its own technology accessible to employees with disabilities. IBM had a reputation that was changing and dependent on the perspective from which it was viewed, giving the company incentive to control their public perception where possible. For example, while IBM computers were wildly successful among businesspeople, Apple's critical comparison between IBM and Big Brother struck a chord with the type of people who constituted Apple's customer base. Apple came into the personal computer market as a new company, with strong ties to the counterculture and an aura of technological user-friendliness. For Apple and its customers, the computer acted as a symbol of individualism and freedom, values that Apple embedded into its machines in a bounded way, prescribing certain options for the user and blocking others. Finally, IBM was geographically diverse, with different parts of the company located across the globe, while Apple had a single headquarters in the Bay Area. These geographical and cultural differences affected how the companies related to external disability organizations and how they organized their own efforts internally to develop and promote accessible technologies.

During the late 1980s and early 1990s, disability and technology activist groups, such as the ATA, sought to transmit knowledge of the potential of personal computer technology to people with disabilities, disability professionals, educators, and legislators. Apple Computer formed the ATA to create a national network of local disability and technology organizations so that their expertise and resources could be more effectively pooled and distributed. The ATA functioned as a bridge between developers and computer users with disabilities to disseminate knowledge and communicate needs. The Alliance demonstrated two levels at which such a social technology could operate. At the national level, the ATA brought organizations together and ran large-scale programs. At the local level, member centers, such as the DCCG, directly worked with individual people to find solutions to their problems. While the ATA grew, leaving Apple to become an independent nonprofit organization, IBM also promoted accessible personal computer technology by connecting with advocacy groups to help place computers into the hands of users and by developing its own accessible technologies for employees and consumers. At the same time that the ATA and IBM increased their accessibility operations, the disability rights movement also regained strength, leading to the passage of federal legislation that more forcefully secured civil rights protections for people with disabilities.

Chapter Four

The Growth of Disability Rights and Accessible Computer Technologies

After diminishing in strength since the mid-1970s, the disability rights movement experienced a resurgence beginning in the late 1980s. This regrowth propelled the passage of new disability rights legislation that affected the use and development of accessible computer technologies, culminating in passage of the Americans with Disabilities Act of 1990. The ADA guaranteed people with disabilities protection from discrimination in a far broader and more enforceable way than previous legislation. As a part of this greater protection of civil rights, technological accommodations were again the means through which people with disabilities could enjoy protection from discrimination and fully participate in society. As with earlier legislation, these new laws offered a public solution to the social problem of disability, in which equal access to social participation was mandated and partially regulated by the government. Rather than requiring individuals to solve their own problems of social access or leaving it purely up to the market to produce accessibility, the laws embraced disability as a social problem whose solutions had to be enforced for everyone. Disability and technology advocacy groups, such as the Alliance for Technology Access (ATA) and its member centers, took part in arguing for the passage of this new legislation and benefited from its enactment, as they moved on to bigger projects reaching more people with disabilities and connecting them with computer technology that could help them.

In this chapter, I situate attempts to improve accessible personal computer technology within the larger context of the disability rights movement and its resurgence in the late 1980s. I begin with a defining moment that advanced the cause of disability rights and placed it firmly in the public eye: the student protests at Gallaudet University fighting for the selection of a deaf president. I examine the roles played by advocacy groups in the passage of two important pieces of federal legislation, the Technology-Related Assistance for Individuals with Disabilities Act of 1988 (Tech Act) and the Americans with Disabilities Act of 1990 (ADA). I then turn to the effects the ADA had on groups like the ATA as new attention and visibility enabled them to turn to larger projects than before. I discuss the ATA's relationship with the computer industry in the early 1990s, as IBM replaced Apple Computer as their main corporate supporter. Comparing the work done on accessibility at IBM and Apple, I look at the different roles they played in philanthropic efforts and the development of consumer products for people with disabilities.

The "Deaf President Now" Protest at Gallaudet

In the late 1980s, the disability rights movement, which had been mostly stagnant for a decade, experienced a national resurgence through public protests and federal legislation. A growing sense of identity helped to urge this movement forward, as groups of people with common disabilities and similar struggles joined together to fight for greater equality. One of the defining events of the disability rights movement that garnered national attention and built momentum for the eventual passage of legislation such as the ADA was the "Deaf President Now" protest at Gallaudet University in 1988. There were a number of reasons that this protest—which demanded the first deaf president in Gallaudet's history—occurred at this time. The population of deaf people in the United States skyrocketed in the 1960s after a rubella outbreak. Medical advances allowed the patients to survive but with hearing loss, an outcome that led to the doubling of the number of deaf children in the country.[1] These children, and all other children with disabilities, would go on to benefit from the 1976 passage of the Education for All Handicapped Children Act and graduate from high school with a higher standard of education than people with disabilities previously experienced. This population of students was college aged in the 1980s, and their high expectations and large numbers gave their demands for equal rights a louder voice. At the same, Deaf activism grew, fighting to preserve Deaf culture and its use of American Sign Language for communication.[2] Additionally, the cumulative effects of 1970s disability rights legislation that granted civil rights protections to all people with disabilities, though often unenforceable, made the fulfillment of new demands a real possibility.

The weeklong Gallaudet protests in Washington, D.C., were of marked importance for the disability rights movement. Unlike the protests over the Section 504
legislation in the 1970s, which were scattered across the country and did not receive national recognition, the Gallaudet protests were covered by the national media.[3] Students demanded that the 124-year-old university (the only university in the
world specifically for the deaf and hard of hearing) appoint its first-ever deaf president. In December 1987, the previous Gallaudet president, Jerry Lee, stepped down,
and a group of Gallaudet alumni planned for a rally on March 1, 1988, to bring the
students, faculty, and staff together in encouraging the administration to choose
a deaf successor. The fliers for the rally explained their position: "With a deaf
person in the position of leadership, one that has the same views, experiences,
and needs that we do, people will become more informed of the needs of deaf
people."[4] A number of politicians and public figures backed the protesters: Vice
President George H. W. Bush, Senators Bob Dole (Republican, Kansas), Bob Gra-
ham (Democrat, Florida), Tom Harkin (Democrat, Iowa), Paul Simon (Democrat,
Illinois), and Lowell Weicker (Republican, Connecticut), Congresswoman Patricia
Schroeder (Democrat, Colorado), and Reverend Jesse Jackson, all of whom sent
letters to the Gallaudet board of trustees in early March supporting the choice of
a deaf president for the federally funded university.[5] Simon, in particular, clari-
fied why having a deaf president for Gallaudet mattered: "A fundamental requirement, overriding any other for this job, is an understanding of deafness—what it
is and how it affects the educational experience."[6] The students of Gallaudet
insisted on being represented by someone who was like them.

The university announced the three finalists for the vacant position the day of
the rally: Dr. I. King Jordan (the deaf dean of the college of arts and sciences), Dr.
Harvey Corson (the deaf president of a school in Louisiana), and Dr. Elisabeth Zinser (an administrator at the University of North Carolina, who was the only hearing candidate and did not know sign language).[7] On March 6, the board of trustees declared that they had chosen Zinser. Mass student protests broke out
immediately following the news, and the following day students blocked all campus
entrances with hot-wired cars and buses, closing down the university. The Gallaudet students, faculty, and staff issued a vote of no confidence in the board of
trustees, calling for the appointment of one of the deaf finalist candidates, the res-
ignation of board chairwoman, Jane Spilman, the alteration of board bylaws to require a majority deaf representation on the board, and no reprisals against those
involved with the protest.[8] Spilman had become a target of the protester's anger
not only for announcing Zinser's appointment but for being quoted as saying, "Deaf

people are not ready to function in a hearing world." Though she denied ever saying it, the quote was picked up and printed repeatedly during the national press coverage of the protests.[9] Classes technically resumed the following day, although 90 percent of students boycotted and continued protesting. Students at other deaf schools across the country and Gallaudet alumni joined the protest.[10] Protesters prevented Zinser from coming to campus or speaking to the student body, which refused to legitimize her authority as president.

The protesters approached the federal government for support and won. On March 9, the front page of the *Washington Post* declared congressional support for a deaf Gallaudet president. Representative David Bonior (Democrat, Michigan), a Gallaudet board member, commented on what might happen if Zinser stayed as president, "I'm concerned it has the potential to hurt the funding of the university, especially when you have leaders from both parties going out of their way to express themselves on this."[11] That evening, on ABC's *Nightline,* Ted Koppel interviewed Zinser and student body president Greg Hlibok. The following day, Dr. I. King Jordan retracted his previous support of Zinser's appointment as president. Later that day, Zinser publicly resigned. The protest continued, including a march to the Capitol, until March 13, when the rest of the Deaf President Now protest's demands were met. Spilman resigned, and Jordan was named Gallaudet University's first deaf president.[12]

One of the reasons for the Gallaudet students' success was the national attention the protest garnered. Across the country, American citizens watched a group of young people with disabilities stand up for their rights and demand representation by one of their own. While the American public viewed this protest as being conducted by a group of people with disabilities and thus, in a way, representing the population of people with disabilities as a whole, the fact that it was by deaf people in particular was significant. Deaf activists distinguished themselves from people with disabilities by arguing that deafness was a culture, with its own language and means of communication, not a medical condition. The Deaf movement historically distanced itself from disability rights; however, from the perspective of the larger society, deaf people faced the same discrimination as people with disabilities, needed technological accommodations in order to fully participate in society, and were part of the same struggle for civil rights and equal opportunities.[13] Congressman Tony Coelho described Gallaudet as a catalyst for the disability rights movement: "It is time, I think, to stand up. I think Gallaudet proved that and sort of lit a spark not only with the hearing disabled but with the disability community all over the country. We do not want to be patient anymore."[14] The outcome of the

Gallaudet protests spurred the disability rights movement to continued action and gained public attention. A few months later lawmakers introduced before Congress the Technology-Related Assistance for Individuals with Disabilities Act and the first version of the Americans with Disabilities Act.

Technology-Related Assistance for Individuals with Disabilities Act of 1988

As the disability rights movement surged forward, newly energized by a population of college-aged adults with disabilities who had grown up reaping the benefits of disability rights legislation from the 1970s, activists and proponents within the federal government pushed for the passage of two significant pieces of legislation: the Technology-Related Assistance for Individuals with Disabilities Act of 1988 (Tech Act) and the Americans with Disabilities Act of 1990. The Tech Act is not included in most histories of the disability rights movement. It was smaller and more specifically focused than the ADA, dealing only with assistive technology for people with disabilities. It offered grants to states to get technology to people who might need it. The Tech Act may also receive less notice because it was uncontroversial and passed through Congress quickly and with bipartisan support. However, it was a vital piece of legislation for disability-and-technology advocates, who frequently wrote about it in their own materials. It provided federal funding to programs that helped connect people with disabilities and assistive technologies to benefit them. It also allowed groups, such as the ATA, to work directly with government agencies and receive funding for larger projects. The Tech Act promoted accessible technology through its system of dissemination, providing resources and information for organizations that worked with users.

In Congress, those who supported the Tech Act were usually people who had personal connections with disability rights. Tom Harkin, the chairman of the Senate Subcommittee on the Handicapped, introduced the bill. Harkin, whose brother was deaf, was a major proponent of disability rights and the only U.S. senator proficient in American Sign Language. He described the passage of the Tech Act: "Following two days of testimony on how technology has already helped the disabled to lead productive lives, it became clear that America needs a comprehensive, responsive, and coordinated system to stimulate new developments and make them accessible and affordable to disabled people."[15] The Tech Act created this new system of technological development and accessibility by encouraging and funding programs at the state level that provided assistive technology and training for people with disabilities. The bill defined assistive technology as any de-

vice "that is used to increase, maintain, or improve functional capabilities of individuals with disabilities."[16] These devices included off-the-shelf and customized
technologies.

Disability rights legislation passed during the 1960s and 1970s made the argument that technology provides access to social participation for people with disabilities and that technological accommodations are necessary for equal opportunities in society. The Tech Act followed in the footsteps of these earlier laws and
was based on findings that technology was a necessary part of people's lives and,
in particular, enabled people with disabilities to

(A) have greater control over their own lives;
(B) participate in and contribute more fully to activities in their home, school,
     and work environments, and in their communities;
(C) interact to a greater extent with nondisabled individuals; and
(D) otherwise benefit from opportunities that are taken for granted by individu-
     als who do not have disabilities.[17]

Beyond the benefits assistive technology could impart to people with disabilities,
the Tech Act also acknowledged that there was an economic benefit to individuals with disabilities and to society as a whole. In the bill, legislators argued that
the use of assistive technology would reduce the cost of social activities such as education, health care, transportation, and telecommunications for individuals with
disabilities, their families, and society. This finding was not backed up with direct
evidence, but the argument seems to have been that it would be cheaper for people with disabilities to take part in various parts of society if they had effective technology rather than ineffective technology or none at all. In an article published
after the bill was passed, Harkin described the cost savings: "Today, the federal
government funds hundreds of millions of dollars in unemployment disability payments to persons who could be employed if they had access to assistive technology. Investments in technology to keep people working can save taxpayers and employers much of the cost of long-term disability payments."[18] At a time when there
was a workforce shortage in the United States, the promise of the Tech Act leading to more employable people offered a concrete, economic benefit to the nation
as a whole and offset the costs of implementing the bill.

The arguments made in the Tech Act marked another step in the slow shift in
how American society viewed people with disabilities. No longer were they regarded as the "deserving poor" who needed charity to survive. Assistive technology could enable people to more fully participate in society, bridging the social

divide between people with disabilities and those without. At the same time, how-ever, the language of the Tech Act differs from that used in earlier rehabilitation legislation, which necessitated the funding of programs that would make people with disabilities employable and thus pay back some of the welfare cost society bore to help them. While the cost-saving arguments in the Tech Act do concern employment and the reduction of disability payments to people who could work if they had access to technology, the bill does not speak in terms of needing to rehabilitate or fix the individual in order for them to be employable. Instead, the Tech Act locates the problem in the lack of available assistive technology and the money that is wasted on welfare when it could be more efficiently spent on providing people with the means of employment; solving the problem in-volves funding programs that will distribute technology to people who are blocked from social participation without it. In addition, the Tech Act focuses on a full range of social activities beyond employment and insists on the neces-sity of assistive technology in allowing people with disabilities to experience fuller lives.

The legislators behind the Tech Act recognized the positive effects that already existing technology could have on people's lives and also sought to fulfill needs for technology that had not yet been met. In arguments similar to those made by the DCCG and NSEA, the Tech Act talked about how a lack of knowledge prevented people from accessing technology, and it also pointed to issues of cost and govern-ment coordination:

There is a lack of—

(A) resources to pay for such devices and services;
(B) trained personnel to provide such devices and services and to assist individuals with disabilities to use such devices and services;
(C) information about the potential of technology available to individuals with disabilities, the families or representatives of individuals with disabilities, individuals who work for public agencies and private entities that have contact with individuals with disabilities (including insurers), employers, and other appropriate individuals;
(D) coordination among existing State human services programs, and among such programs and private agencies, particularly with respect to transitions between such programs and agencies; and
(E) capacity of such programs to provide the necessary technology-related assistance.[19]

This argument describes the lack of a social technology to communicate information on assistive technology to users. The Tech Act addressed the need for a social technology that operated on a far larger scale than even an umbrella disability and technology advocacy network like the ATA could. What was needed was a way to coordinate communication efforts across the entire country and include government agencies, individuals with disabilities, and people with technical knowledge. This problem of lack of access to technology was compounded by what the Tech Act described as a lack of motivation for technology companies to develop products aimed at consumers with disabilities because the companies perceived a limited market. The bill also explained that federal agencies lacked the coordination to provide for assistive technology. The solution created by the Tech Act was to fund state programs that would increase awareness of the technological needs of people with disabilities, increase technological knowledge for those people and those close to them, explore procedures that were either providing for assistive technology or blocking access to it, coordinate state agencies and private entities to provide technology, and, overall, increase the opportunities for people with disabilities to access assistive technology. At the federal level, the Tech Act also included efforts to uncover policies that enabled or impeded funding of assistive technology, to remove obstacles to funding, and to improve the federal government's ability to supply the states with assistance in providing assistive technology.[20]

The Tech Act was not strictly top-down legislation created by politicians; disability activists played a role in enacting this new policy. The ATA was involved with the Tech Act before its passage as well as after the federal government began doling out grants to state programs. In the spring of 1988, the Alliance Planning Team, through its connection at the time with Apple Computer, submitted written testimony to Congress about the Tech Act prior to its passage.[21] Most of Apple's concerns dealt with the need to achieve equity and the possibility of doing so through making adaptive technologies available. Apple and the Alliance argued that legislation was needed to make technology more available, especially to those people who fell through the gaps between service providers: "Concern for equity cuts across many of these questions and is a central issue in barrier-free technology. Often, the people who should benefit most from adaptive technology are the people who can least afford it. Many children and adults with disabilities are blocked from accessing . . . technology in their communities because they belong to the wrong age group, disability group, socioeconomic group or educational services group." Any federal legislation that provided assistive technology needed to find a way to reach those people who had difficulty affording it or were left out of current technology distribution channels. From Apple's perspective, the Tech Act needed to provide technology as a way to work toward equity in society for people with disabilities: "We firmly believe that a program which provides loaned, free or reduced priced equipment; assists consumers in seeking public and private funding; or enables individuals with disabilities to qualify for a low cost or subsidized loan program is necessary for equity and should be a substantial part of this legislation."[22] The existence of assistive technology itself was again not enough; though the focus here is on monetary cost, not lack of information, technology still needed to be better made available to people with disabilities who could benefit from it.

To give the federal legislation a model to emulate, Apple argued that the NSEA offered a positive example of an organizational structure that was capable of reaching people. One of the major reasons for the NSEA's success, according to Apple, was its core partnership between consumers and industry professionals: "We believe that the inter-disciplinary, cooperative approach characteristic of the NSEA is a critical component in any comprehensive adaptive technology legislation. We believe that the NSEA model takes advantage of systems, organizations, and structures that are currently in place, and introduces new technology and information on a daily basis. The model of the NSEA is especially intriguing because it represents both a healthy partnership between the public and private sectors and a community-based, collaborative approach for getting everybody to work together." The NSEA's network enabled it to take advantage of systems of expertise already in place and connect them together to better share knowledge; its network allowed it to function on multiple levels—from one-on-one community work, to local groups working with each other, to larger, national projects that attempted to reach many people at once. Apple also emphasized, as crucial to success, the roles of consumers with disabilities and the parents of children with disabilities.[23] The company argued that such a model would provide significant strengths to the funding network the Tech Act would construct. The NSEA, in the form of the Alliance for Technology Access, would continue to work with and reap the benefits of this funding network after the passage of the bill.

The Tech Act directly influenced disability and technology advocacy efforts around the country, including those of the newly independent ATA. After the Tech Act passed, the federal government began giving out a certain number of grants to states each year. By 1993, forty-two states had received Tech Act grants. Bob Glass explained that forty-two ATA centers had, at that time, been in contact with the assigned agencies in their states and that nineteen centers were receiving some amount of Tech Act funding through their state agencies.[24] As the Tech Act expanded its coverage across the country each year (by 1995, all fifty states were covered), the ATA responded by including more centers in more states under its purview. In addition, in 1990, the ATA began the ACTION Project (Accessing Computer Technology In Our Neighborhoods), funded by a grant from the U.S. Department of Education under Title II of the Tech Act.[25] The project planned to involve five resource centers over three years and focused on technology for people with low-incidence disabilities.

The Americans with Disabilities Act of 1990

While the Tech Act was a great boon for efforts to promote accessible and assistive technologies for people with disabilities, its success was eclipsed two years later with the passage of more general antidiscrimination legislation, the Americans with Disabilities Act of 1990, arguably the greatest success of the disability rights movement. Lawmakers and activists had worked to develop it throughout the late 1980s. In 1986 the National Council on the Handicapped to the President and Congress published a report called "Toward Independence: An Assessment of Federal Laws and Programs Affecting Persons with Disabilities—with Legislative Recommendations."[26] This report was the result of the Rehabilitation Act Amendments of 1984, which established the National Council on the Handicapped as an independent federal agency, tasked with reviewing the efforts of federal programs related to people with disabilities and recommending ways to improve them.[27] The council was made up of fifteen independent experts in disability issues. One of the main conclusions they reached in their report was that "federal disability programs reflect an overemphasis on income support and an underemphasis of initiatives for equal opportunity, independence, prevention, and self-sufficiency."[28] In contrast to Section 504's limited scope and problems with enforcement, the federal government now recognized the disability rights movement's call for civil rights and access to full participation in society. This recognition marked a significant step toward disability being seen as a social problem that needed to be solved through public efforts to improve accessibility. In the report, the council recommended the enactment of an equal opportunity law on the basis that, "If the goals of independence and access to opportunities for people with disabilities are to be achieved, it is essential that unfair and unnecessary barriers and discrimination not be allowed to block the way."[29] The council argued that existing laws (including Section 504) were inadequate and not broad enough, as compared to antidiscrimination laws for other populations. They called for a law that would make it clear to society as a whole that discrimination against people with disabilities was unacceptable. They suggested that such a law be called the Americans with Disabilities Act and apply to all federal departments, all federally funded programs, all employers with more than fifteen employees, all landlords and providers of housing, all public accommodations, all interstate transportation businesses, all insurance providers, and all state and local government agencies.[30] So as not to repeat the problems with legislation like Section 504 of the Rehabilitation Act, this new law should also have specific enforcement policies.[31]

Two years later, in January 1988, the National Council on the Handicapped followed up their previous report with a new assessment, "On the Threshold of Independence: Progress on Legislative Recommendations from *Toward Independence*."[32] Included in this report was a draft of proposed legislation called the Americans with Disabilities Act of 1988. The council found that in the previous two years, 80 percent of their recommendations had been at least partially accomplished. Twenty-one statutory provisions had been enacted, and a further eight bills had been introduced to Congress that would help accomplish the goals set out in the previous report. Public consciousness toward disability rights also had increased during the previous two years. "Toward Independence" found favor with the disability community and the general public, and the report appeared on national news. With progress being made in many areas, the main recommendation became the "enactment of a clear and comprehensive statute guaranteeing equal opportunities for people with disabilities,"[33] a goal the ADA would come to fulfill.

Lawmakers introduced the first version of the ADA before Congress on April 28, 1988. It was written by Robert L. Burgdorf Jr., a disabled attorney and research specialist for the National Council for the Handicapped.[34] Its sponsors in the House of Representatives and Senate were Tony Coelho (Democrat, California) and Lowell Weicker (Republican, Connecticut), respectively. Both men had experience with the discrimination faced by people with disabilities, Coelho as a man with epilepsy and Weicker as the father of children with disabilities.[35] A joint hearing of the proposed bill was held before House and Senate subcommittees on September 27, 1988.[36] Present were Coelho and Weicker, as well as Senators Tom Harkin and Edward Kennedy and Representatives Major Owens, Matthew Martinez, and James Jeffords. Witnesses gave expert testimony on the subject of discrimination toward people with disabilities, including an account of the Gallaudet Deaf President Now protests by Greg Hlibok.

The efforts of all the people fighting for civil rights protections for people with disabilities culminated with a law that would change the way people with disabilities fit into American society. Congressman Owens (Democrat, New York) described the change: "This legislation grants full rights to Americans with disabilities and moves our great Nation from a respectable position of official compassion for those with impairments to a more laudable position of empowering disabled Americans."[37] The concern with empowerment echoed disability rights legislation from the 1970s, such as Section 504. Empowerment marks a turn firmly away from the previous medical model of disability legislation Owens went on to give credit for development of the ADA to the disability rights movement and cited the Gallaudet protests as having made the movement "highly visible."[38] Coelho explained how the Gallaudet protests and the disability rights movement had affected him personally: "What happened at Gallaudet University was an inspiration to all of us with disabilities, in that if we ourselves believe in ourselves and are willing to stand up we can make a difference. That is what this bill is all about; 36 million Americans deciding it is time for us to stand up for ourselves, to make a difference, to say that we want our basic civil rights also. We deserve it."[39] Judy Heumann, the leader of the Independent Living Movement for whom Jackie Brand worked at the Center for Independent Living, gave testimony expanding upon why it was time for legislation such as the ADA to come to pass: "I personally think that the Gallaudet experience and the 1977 demonstrations in relationship to 504 and the subsequent Development of Independent Living centers and community-based organizations around the United States, and the real true emergency of a rights movement are going to compel the United States to recognize its responsibility."[40] These statements demonstrated one of the strengths behind the disability rights movement that made it powerful enough by the late 1980s to have legislation like the ADA in consideration before Congress. The ADA would cover one out of seven Americans when it passed in 1990; the sheer ubiquity of people with disabilities made them a population that had a presence everywhere.[41]

This strength of numbers was not enough to pass the first version of the ADA. Part of the issue was timing. The joint hearing on the ADA took place less than a month before the end of the 100th congressional term. There was no press coverage of the 1988 bill, elections were upcoming, and the Reagan administration was winding down and distracted by other issues.[42] The National Council on Disability (the successor to the National Council for the Handicapped) produced a history of the development of the ADA that describes a strategy to take advantage of the coming presidential election by soliciting the candidates' support for the ADA while they were competing with each other.[43] This approach worked particularly well with then vice president George H. W. Bush, who gave repeated public support for people with disabilities and courted their votes during his campaign.[44] Harkin explained during the 1988 joint hearing that progress would not be made on the bill that year and that the intention was to reintroduce the ADA the following year.[45] Although the 101st Congress did eventually pass the ADA, the entire process of negotiation and rewriting lasted until their second session in 1990, and it was an altered version of the bill that passed, with new champions behind it.

The large population of people with disabilities played a major role in facilitating the bill's passage. As people with disabilities experienced a growing awareness of their group identity and common struggle for equal rights, disability activists across the country and in Washington mobilized them to defend the bill. Two of the main strategists trying to influence the government were Pat Wright (Judy Heumann's assistant during the 1970s San Francisco Section 504 protest) and Ralph Neas (a prominent civil rights attorney).[46] The ADA brought together disability rights advocates with broader civil rights advocates to join forces in securing enforceable civil rights legislation for people with disabilities. Within the federal government, many of those involved with the ADA were either themselves disabled or had close family members with disabilities.[47] Even the newly elected President Bush had experience with disabilities in his family, having a daughter who had died in infancy of cancer and a son who had learning disabilities.[48]

In spite of this support, the final version of the ADA that passed in 1990 only succeeded because of the changes made to it. The first version had been more radical and was seen by Harkin and Kennedy as having little chance of passing.[49] It stipulated that all buildings and public transportation vehicles had to be made accessible, and the only exceptions were interpreted as meaning that a business would only be allowed not to make accessibility improvements if doing so would bring it to the brink of bankruptcy. The final ADA brought back the Section 504 language, granting businesses exceptions in cases of "undue hardship," which was to be interpreted on a case-by-case basis.[50] The final ADA also only required the removal of barriers for new buildings and vehicles, and it mandated the altering of existing structures only if accessibility was "readily achievable." If it was not, then alternative services had to be provided for people with disabilities.[51] Other major changes included limitations on legal actions available in discrimination cases and an overall change in tone away from emphasizing the intolerability of discrimination and toward encouraging more proactive methods of meeting accessibility standards.[52] The ADA passed the Senate 76 to 8 in late 1989, but House negotiations lasted until May 1990, when it finally passed 403 to 20. President Bush signed the ADA into law on July 26, 1990, in front of three thousand people gathered on the White House lawn. It was the most-attended bill signing in U.S. history.[53]

The ADA was a general antidiscrimination bill, but in order for it to be enacted technology would have to play a vital role. While the legislation was a public attempt to tackle the problem of disability, it relied on accessible technologies developed by corporations. In an ATA publication that came out shortly after the bill was passed, the organization explained its position and desire to work with the ADA: "Technology is going to play a leading role in the realization of ADA. The Alliance is committed to insuring that the implementation of ADA is not impeded by the lack of awareness and information about the potential of technology to make equality a reality. Working locally and nationally with planners and employers, the Alliance has an important role to play in supporting our new 'Declaration of Independence.'"[54] According to the ATA, equal opportunity for people with disabilities would only be possible in our society via technology. The ATA's view here recalls some of the ideas behind legislation such as the Architectural Barriers Act of 1968, such as the idea that society has been constructed with barriers preventing people with disabilities from full participation. Technology allows for a means of overcoming barriers, providing accommodations that different people need in order to access society and interact with other people. Technology, then, is the tool through which civil rights are made attainable for people with disabilities; even if the social environment was designed all along for deliberate universal access and people with disabilities as intended participants, bodies present limitations that technology can accommodate. Equal participation for everyone is only possible if differences in bodies are understood and accommodated.

The Alliance for Technology Access after the ADA

After the passage of the ADA, the Alliance for Technology Access continued to grow and expand through the early 1990s. Bob Glass estimated that in 1991 alone, the ATA and its centers provided services for around 72,000 individuals and had over 1,000 people with disabilities, parents, and disability professionals in leadership and advisory positions.[55] The network structure of the ATA changed as the organization grew. As a national alliance connecting small, discrete groups under one umbrella, the ATA began to form new networks in the form of large-scale national projects. These projects offered different ways of attempting to connect people with disabilities and computer technology that might aid them. Celebrating its national involvement in such projects, the ATA commemorated its five-year anniversary with a photo collage forming a map of the United States, showing off its work across the country and the different people it helped.

One of the most prominent ATA projects begun in the early 1990s was Compu-CID (Computer Classroom Integration Demonstration), a three-year, federally funded project, which dealt with the use of computers in supporting the mainstreaming integration of students with disabilities into classrooms with nondisabled students.[56] CompuCID was an important project for the ATA at this time. Mainstreaming efforts were occurring simultaneously with the development of personal computers, so computers and students with disabilities were entering mainstream classrooms at the same time. The introduction of computers into schools was particularly meaningful for those students who would depend on them as necessary tools that enabled them to participate in mainstream classes. Educators needed to be trained to integrate into their classes the new students, whose needs they might be unfamiliar with, and the new technology, about which they might know little. Groups like the ATA were poised to instruct educators in the best ways to integrate these children and machines into the classroom.

The CompuCID project involved ATA resource centers working with six public school districts in Colorado, North Carolina, California, Tennessee, and Washington. In each location, the project was run by a technology team made up of a local educator and a person with a disability (or the parent of a child with a disability), who both had computer expertise. Using methods such as cooperative learning and cross-age tutoring, the project attempted to change how students were being taught as well as how technology was used in the classroom.

The different sites followed loose guidelines as part of the project. Computers had to play a role in integrating disabled and nondisabled children, and facilitators had to try experiments in cooperative learning (with students working in groups together toward educational goals). The classroom circumstances at the different sites varied widely, however, and were taken into account in how the technology teams tackled demographic and cultural differences. One of the California classes involved a mix of children for whom English was a second language, children with learning disabilities, and one child with severe physical disabilities. The school in Colorado practiced team teaching, whereas the North Carolina program attempted to use computer software to address improvements in basic skill levels.[57] Teachers who took part in CompuCID found that, as intended, computer technology played an essential role in classroom integration. A CompuCID newsletter reported on the experiences in one classroom in late 1990: "Beth Pitts, a third grade teacher in one of the demonstration classrooms in North Carolina's Cornelius Elementary School, said she has seen the computers serve as a common bond for different types of students in her classroom. 'It's been very successful because they (students with disabilities) can do as well as anyone else does in the classroom. The computer puts them at equal.'"[58] The technological accommodations the computer provided allowed these children to be perceived as equals in the classroom. Pitts's comments about computer technology helping to level the playing field for children echoed the ATA's belief that the computer was a universalizing technology providing new forms of communication that could change the meaning of disability. The computer was a tool that, unlike other traditional education tools that people with certain disabilities would be unable to operate, all students could use once it was made accessible. Both Pitts and a teacher at the Colorado site went on to praise computer activities in promoting teamwork and a sense of community among the students.

In order to ensure that people with any type of disability could access computer technology—and thus move more toward true universal access, in which every type of use is considered and accommodated—the ATA began a project focused on uncommon disabilities in 1990. Funded by a grant under Title II of the Tech Act, the ACTION (Accessing Computer Technology in Our Neighborhoods) project aimed to teach people with low-incidence disabilities about assistive technology that might benefit them. The Alliance developed and tested a model for outreach and technological training, with the goal of showing how computers could improve social integration and independence for people with less common disabilities who had not yet had opportunities to learn about computer technology. The Alliance used its local resource centers to find ways to connect with people in those communities. The project held technology demonstrations in heavily visited public areas, conducted individual and small-group training sessions, produced videos of people with low-incidence disabilities using technology (to be used by the individuals themselves and to be shared with others), and taught individuals with disabilities and their families about assistive technology, funding methods, and relevant legislation.[59] The ACTION Project finished in September 1994.[60] After its conclusion, the ATA developed a manual on outreach methods.[61] Projects (such as ACTION) that explicitly address the needs of people with low-incidence disabilities are necessary if the ideals behind universal design are to work. In order to create technology that can be used by everyone, even uncommon requirements of use need to be addressed.

After the passage of the ADA, the Alliance continued to grow, both in the scale of projects it could direct and in renown as an advocacy group. As the ATA expanded, it also gained increasing national recognition. Starting in 1991, the Alliance established an honorary board of directors made up of prominent disability activists and people with disabilities, including Stephen Hawking, Christopher Burke (an actor with Down Syndrome well-known for his role on the TV series *Life Goes On*), Sandra Parrino (the director of the National Council on Disabilities), Johnny Wilder (a quadriplegic jazz musician), Max Schliefer (editor of the *Exceptional Parent*), and Judy Heumann.[62] Heumann had worked with Jackie Brand, and Wilder and Schliefer also had close personal ties to the ATA. In interviews, Alan Brightman often relates a story about the keynote address at an American Occupational Therapy Association conference during which Wilder demonstrated his ability to write music on a Macintosh computer using a sip and puff straw switch, which allowed a user to control a computer by blowing into a straw. Brightman shared this anecdote as an example of how rehabilitation professionals needed to be convinced of the power of the personal computer in providing access to activities many people believed impossible for individuals with disabilities.[63] Schliefer's *Exceptional Parent* magazine for parents of children with disabilities regularly published articles on the DCCG and ATA.

The ATA was not capable of unlimited growth. There were limitations built into its network structure that kept the Alliance stabilized at around forty resource centers. Being composed of a network of independent, local centers, the ATA depended on these centers to maintain themselves. Sometimes these centers, which, in turn, depended on the energies of the people who operated them, ran out of steam and collapsed. While the ATA had strong successes and steady growth during the first five years after its founding, there were also failures; six resource centers lost Alliance membership during this time. Of those six, one was shut down suddenly by its own umbrella organization, two closed after the people running them left and no one else took over, and three were removed from the Alliance after they failed to meet minimum standards of operation and would not improve after the ATA attempted to help.[64] In response, the ATA developed a list of potential indicators that a center might be falling apart. With enough warning, the larger organization could step in to provide assistance. According to Bob Glass, these indicators included: "Little or no presence on AppleLink; difficulties between the center and a sponsoring, dominant umbrella organization; strong dependence on single individual or couple who leave the center or community; failure to return the annual Program Impact questionnaire; and failure to send at least one representative to a national training event."[65] When at least two of these conditions were present, the ATA would offer assistance to the struggling resource center.

The ATA and IBM Partnership

While the ATA may have been limited in how far it could grow with its local resource centers, the Alliance moved on to bigger projects by combining its efforts with those of corporations beyond Apple Computer. The ATA was not on its own in maintaining its projects and resource centers; corporate involvement and donations continued after Apple formally disconnected itself from the Alliance. By 1991, IBM had stepped in and replaced Apple as the new major corporate sponsor of the ATA.[66] IBM's partnership with the ATA was not as close as Apple's had been but was just as generous in terms of donations.

Whereas Apple had partnered with groups it encountered locally, IBM, a global company with offices in many places, worked with numerous disability organizations that were nationally focused or with widespread local branches. IBM's work with the Easter Seals to provide computers at a discount to people with disabilities is one example. During the 1990s, the National Federation of the Blind (NFB) maintained a particularly warm relationship with Jim Thatcher, the IBM researcher who worked on the development of the Screen Reader. Curtis Chong, the president of the NFB in computer science, applauded Thatcher's work on the Screen Reader and his willingness to present at NFB annual meetings.[67] In 1993, the NFB presented IBM with a letter of support for its development of the Screen Reader/2 for OS/2 and its graphical interface.[68] Other national organizations that IBM partnered with through their local branches were the United Cerebral Palsy Association and the Alliance for Technology Access.

In the early 1990s, IBM partnered with the Alliance for Technology Access as one of their major corporate supporters. There were two aspects to this relationship; IBM provided technological resources to Alliance disability and technology resource centers across the country, and it participated in large-scale projects with the ATA. IBM began appearing as a major contributor in ATA annual reports during this time. The ATA described its appreciation for IBM's generosity: "IBM merits immense respect in the field of assistive technology, and FTA is both pleased and proud to be partners with IBM in the promise of technology."[69] In 1991, IBM began a long-term loan of $250,000 worth of software, adaptive devices, and technical support to ATA resource centers. The ATA worked alongside IBM's National Support Center for Persons with Disabilities to install IBM computers with Independence Series products (at this time, Screen Reader, SpeechViewer, and Phone-Communicator) at the centers. In addition, IBM provided a suite of educational software in reading, language, math, science, and typing.[70]

IBM also played a part in a national project that intended to teach computer skills to children and to treat children as intended computer users; as a part of this project, children with disabilities were explicitly acknowledged as computer users. In 1991, IBM joined with the Mattel Foundation (the toy company's nonprofit, charitable, offshoot organization that helps children in need) and the ATA in the Computer Learning Lab Project (renamed, in the late 1990s, the Mattel Family Learning Program). The Mattel Foundation had started the project in 1990, installing computer labs with IBM equipment in schools across the country for use by students in kindergarten and first grade. A 1991 article in the *Cherokee Country Herald* described the project and one of the labs that was being set up in a local school.[71] By 1991, Computer Learning Labs were in place in thirty schools, with 1,500 students using the labs. The labs employed IBM's Writing to Read software, a phonemic spelling system that allowed children to write any word they knew before they were old enough to learn proper spelling and grammar. Mattel favored the Writing to Read program over similar programs because of its consistency in how it taught the user. The personal computer, as a universal tool that allows for new forms of learning and communicating, offered possibilities as a tool to teach reading and writing in ways that improved upon traditional educational tools and methods. In addition, each computer station in the labs offered four other pieces of software. Kidware by Mobius allowed younger students to prepare for Writing to Read, and Talking Textwriter by Scholastic was a simple word processor that provided voice feedback. IBM contributed SpeechViewer to help students with speech disabilities and a Spanish-language program for students who were bilingual.[72]

In order to accomplish the goal of building labs that met the needs of all children using the computers—including children with disabilities—Mattel needed the expertise of disability and technology activists. The ATA joined the project to expand its scope to include improving access to computer tools in the classroom for children with disabilities, particularly for children with multiple disabilities.[73] Local ATA resource centers connected with schools hosting Learning Labs and provided training and support for the teachers, children, and their parents. First, the ATA developed training programs to teach educators and parents how to use the computer technologies, particularly the adaptive devices and accessibility features needed by children with disabilities. In 1992, the ATA organized a national training meeting for educators, parents, and ATA staff involved with the labs.[74] The Alliance also worked directly with the Writing to Read software to better permit its use by children with disabilities.[75] Second, the ATA helped organize online functions for the labs, providing further training, technical support, and a Web site for the project.[76]

By 1999, the Computer Learning Labs had engaged 4,150 children with disabilities. IBM does not appear to have stayed involved with the lab project for long after the ATA joined. Mattel and the ATA expanded the program to allow schools to choose the technology that would work best for their individual programs; schools did not only have to use IBM's Writing to Read software. In a lab installed in 1998 at California State University Northridge's Child Development and Family Relations Lab School, Apple Power Macintosh computers were used instead of IBM or IBM-compatible ones.[77] The Computer Learning Lab Project, over the course of the 1990s, succeeded at bringing computers into schools for children to learn how to use, and it also demonstrated some of the potential of the computer as a new kind of tool that provided ways to learn skills such as reading and writing. The project intentionally included children with disabilities as students in the classroom and as computer users, enacting the potential of the computer as a universal technology, usable by and beneficial to everyone.

IBM's work on projects like the Computer Learning Lab and its partnership with the ATA demonstrates some of the diverse ways that major computer companies did work with and for people with disabilities during the 1980s and 1990s. The development of accessible personal computer technologies and interactions with disability and technology activist organizations reflected the different values and histories of IBM and Apple, two large-scale, general computer companies. IBM's methods of working with people with disabilities were very internally organized. The company created multiple accessibility features to benefit its own employees, IBM employees with disabilities developed technologies and projects that might aid themselves and other people with disabilities, and IBM established programs to train people with disabilities in computer-related careers. At both IBM and Apple, much of the impetus to focus corporate attention on the accessibility of personal computers during the 1980s came, in large part, from a single nondisabled employee. Jim Thatcher and Alan Brightman showed interest in computer technologies that could benefit people with disabilities and worked within their corporate environments to bring attention to accessibility needs. Brightman founded Apple's Office of Special Education and Rehabilitation, while Thatcher's work creating the IBM Screen Reader and the positive reaction to it from the blind community led to the formation of IBM's Independence Series of products for people with disabilities and its organizational division of accessibility work between the National Support Center, Special Needs Programs, and Special Needs Systems.

Unlike Apple, IBM more publicly promoted its disability endeavors. IBM's longer history and strong public opinions, both positive and negative, may have led to a greater need for the company to publicize its accessibility work, hoping to influence public opinion during and after the Justice Department lawsuit. IBM touted its diverse hiring practices and training programs for people with disabilities before the advent of the personal computer. During the 1990s, when both Apple and IBM experienced near catastrophic losses, Apple chose to discontinue its internal accessibility group in a cost-savings move. IBM, however, kept its accessibility groups, and efforts to develop accessibility features continued during its downturn. Finally, IBM avoided much of the public criticism over the lack of accessibility of the graphical user interface that was directed at Apple and Microsoft during the late 1980s and early 1990s. IBM developed its own in-house screen reading technology instead of relying on third-party developers and the need to provide them with documentation and access to the operating system. IBM's Screen Reader/2 was the only screen reader option for OS/2 (an operating system whose success was short-lived), but IBM's quick work in developing the software drew praise for the company from the National Federation of the Blind. At the same time the NSB was harshly criticizing Microsoft. IBM's fundamental values did not include the user-friendliness, design aesthetic, or utopian possibilities of computer technology that lay at Apple's core; instead, IBM followed goals of diversity and market domination within a large, complex corporate structure that allowed for small, personal projects to thrive and become consumer products.

As the children who experienced the benefits of 1970s civil rights legislation grew up and became adults, they propelled the disability rights movement forward with a new sense of identity. Protests at Gallaudet University garnered national attention and growing momentum toward stronger civil rights protections for people with disabilities. The passage of the Tech Act and Americans with Disabilities Act within two years of each other provided better, more enforceable antidiscrimination protection; in order for the civil rights they guaranteed to come to fruition, however, technological accommodations were necessary to enable people with disabilities to fully participate in society. The Alliance for Technology Access took part in fighting for the passage of these laws and, after their passage, in utilizing the resources the legislation provided in developing larger, more inclusive projects across the country to connect people with technology. These accomplishments in disability rights paid off in the 1990s with greater national awareness of the need for accessibility, as well as the increasing involvement of major computer companies in disability and technology activism. IBM, in particular, stepped into Apple's shoes as the main supporter of the ATA, providing resources to the Alliance and taking part in projects with them. As computer technology improved, however, it would also create new barriers for people with disabilities that would need to be overcome.

Chapter Five

Accessibility and Software Applications in the 1990s

A ccessible input and output technology (such as adaptive devices, speech-to-text hardware and software, and screen readers) allowed people with disabilities to have physical access to a personal computer, but they still needed to be able to use software applications on the machine. In the 1990s, software and operating systems became the main focus for disability advocates. For the most part, their efforts focused on the same software everyone else used: word processors, spreadsheets, graphics programs, games, e-mail, and Internet browsers. Ideals of designing buildings and technology to work for all users coalesced into the concept of universal design, culminating in 1997 with "The Principles of Universal Design."[1] The development of accessible software from the 1980s through the 1990s reflected the computer industry's acceptance of these values.

Many of the accessibility features built into operating systems in the 1980s worked with different software applications, allowing people with disabilities to control their software in the ways they needed. These features included, among others, screen enlargement or zooming, disabled repeat keys, and disabled multi-key presses. These built-in features did not work with all software equally well, however. The values behind universal design, which were increasingly being taken up by technology developers, provided a solution to the problem of making software work for all users. Certain third-party software vendors addressed issues of accessibility by developing their applications with these ideals in mind, looking to provide ways for accessible technologies to work with general ones and to maximize the number of users who could use their technology.

Instantiating the values of what became universal design into the development process was one way for companies to increase their user base in the skyrocketing personal computer software industry. By the mid-1990s, the U.S. software industry brought in more than half a billion dollars annually, and that figure was climbing each year, with Microsoft controlling around half of the market share.[2] The Internet was also growing and becoming more commonly used at this time, and more people were conducting more of their personal business on Web sites. Not only did Internet browsers need to be made accessible but also the Web sites that people would want to visit. Personal computer technology began to stabilize during the mid-1990s, as Microsoft Windows became the dominant operating system and what had previously been radical innovations became standardized. In this chapter, I explore different ways that software developers dealt with accessibility, both from within operating systems and in third-party applications, as personal computer technology became involved with more aspects of everyday life. I also examine the role activist groups continued to play in demanding access to personal computers for people with disabilities. By the end of the 1990s, personal computer accessibility had become less the work of small companies and individuals and more the norm across the computer industry, as technological features allowed more people to use computers the way they needed to.

I begin by discussing large changes that took place in personal computer technology during this time; specifically, I analyze the paradigm shift in computer operating systems that resulted in a switch from a text-based to a graphical user interface. This change significantly affected how people used computers, but it took around a decade to cement itself in the technology. People with different kinds of disabilities and their advocates anticipated the graphical user interface differently; some viewed it positively, while others did not. I delve into the accessibility work done by one major software company, Brøderbund Software, and its partnership with the Alliance for Technology Access in the late 1990s. I then shift gears to discuss the work of the Alliance and the Disabled Children's Computer Group in the mid-1990s, as the organizations grew and changed, taking on projects involving software and the burgeoning Internet. Finally, I conclude my history of the development of personal computer accessibility by looking at the late 1990s, when the ATA underwent a change in leadership and Apple Computer's Worldwide Disability Solutions (formerly the Office of Special Education and Rehabilitation) was dissolved. At the turn of the century, certain battles had clearly been won, but others were continually emerging to challenge the accessibility of personal computers for people with disabilities.

From Text-Based to Graphical User Interface

Perhaps the final, major paradigm shift in personal computer technology to directly affect the users' experience was the gradual domination of the graphical user interface (GUI) over the text-based (or command-line) user interface. This technological change altered all users' interactions with the personal computer and had particular salience for users with certain kinds of disabilities as their needs now were either accommodated more fully or ignored. This innovation led to, for the most part, personal computers being more user-friendly, but, as with any change in a technology, certain assumptions were built in regarding who would use it and how. People with certain kinds of bodies found GUIs an improvement, while others—particularly those with vision impairments—experienced a new obstacle in interacting with the computer. This shift did not occur quickly; although invented during the 1970s, the GUI was not available on commercial personal computers until the mid-1980s and was not the ubiquitous interface until the mid-1990s. This gradual switch allowed computer users to anticipate, both positively and negatively, the change from text to graphics. Proponents of personal computer accessibility criticized the development of the GUI for users whose needs they feared would not be met. However, this long span of time also allowed technological solutions to these problems to be found and put into place, to some extent, as the technology developed.

Until the mid-1980s, personal computers used text-based interfaces only, such as Apple Computer's DOS or Microsoft's MS-DOS. These operating systems required the user to enter commands in the form of text via the keyboard. The computer screen displayed output for the user in the form of characters on lines. Graphics became possible as text-based interfaces developed, but programs had to enter a special graphics mode to display them, and they were limited in terms of realism and detail. These computers could not even display text to look as it would when printed out, as the computer screen only displayed characters at a set size and shape. The reason behind such limitations lay in the way text-based operating systems efficiently used the scarce computer memory available in early personal computers. A display buffer in the computer's memory stored the information that the computer would output on the display in the form of ASCII codes for each character indicating its content and properties, such as bold, underline, or color. As only those characters that needed to be displayed were stored in memory, it was far less resource intensive to only light up those characters being used at any given time; hence, early personal computers displayed only the now-iconic bright green text on a black screen.

Software interfaces were not standardized in text-based operating systems, and the controls for one program might look nothing like the controls for another. A 1990 report from the TRACE Center at the University of Wisconsin–Madison, explained how a user's experience was affected by early text-based interfaces:[3] "With early traditional interfaces, one had no choice but to learn the keyboard commands and procedures for each application used. Commands and file names were typically typed again and again each time they were run or opened. Often, interaction with the machine required a tedious dialog of prompts and typed verbal commands."[4] Not all text-based interfaces were so difficult to use, but software developers were not required to follow standardized rules, and, as text-based interfaces were simple to write, programmers tended to create their own custom interfaces for each application.[5] On the whole, while text-based interfaces were simple to write software for and efficiently used the computer resources available, they could be complicated for users to learn and operate, and also lacked standards and high-resolution graphics.

The graphical user interface would fundamentally change how the computer was controlled and how output was displayed to the user. Although the GUI was developed at Xerox's Palo Alto Research Center (PARC) during the 1970s, it was not available as a consumer technology until Apple released first the Lisa in 1983 and then the Macintosh in 1984. Apple's operating system, System, used a GUI that today looks familiar. Using a desktop or office metaphor with standardized windows, icons, and cursors, the user selected and clicked on graphical representations of operating system commands to navigate and control the computer. The GUI displayed output on the computer screen through bitmapping; the screen was divided into pixels, and output information was translated into groups of pixels that were lightened or darkened. The status of all pixels was always stored in memory. While this display method was more resource-intensive, it allowed for dark text on a light background at no greater cost than light text on a dark background, as well as far greater possibilities for detailed graphics and text that looked the same as it would when printed. With a GUI, the operating system also enforced standardized interfaces for the first time. Each program would run within similar looking windows, with similar menus offered across programs. Software programmers had to follow rules set by the operating system manufacturer in order to utilize operating system tools. A comparison of creating a new folder on the hard disk in MS-DOS 6.0

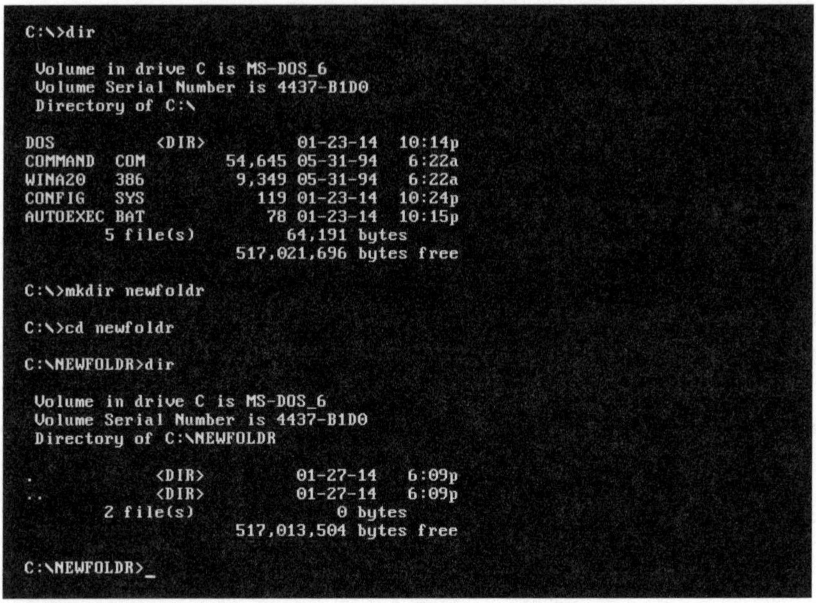

```
C:\>dir

 Volume in drive C is MS-DOS_6
 Volume Serial Number is 4437-B1D0
 Directory of C:\

DOS              <DIR>            01-23-14   10:14p
COMMAND   COM          54,645 05-31-94    6:22a
WINA20    386           9,349 05-31-94    6:22a
CONFIG    SYS             119 01-23-14   10:24p
AUTOEXEC  BAT              78 01-23-14   10:15p
         5 file(s)           64,191 bytes
                        517,021,696 bytes free

C:\>mkdir newfoldr

C:\>cd newfoldr

C:\NEWFOLDR>dir

 Volume in drive C is MS-DOS_6
 Volume Serial Number is 4437-B1D0
 Directory of C:\NEWFOLDR

.                <DIR>            01-27-14    6:09p
..               <DIR>            01-27-14    6:09p
         2 file(s)                0 bytes
                        517,013,504 bytes free

C:\NEWFOLDR>_
```

Figure 5.1. Creating a new folder in MS-DOS 6.0. The simplicity of the text-based operating system made it relatively easy for screen readers to read out loud the text to a blind computer user, but it required all users to memorize commands, and different software applications tended to have different-looking interfaces. Courtesy of Matt McGuire.

and Windows 3.1.1—both released in 1993—demonstrates some of the differences in using personal computers that had text-based versus graphical user interfaces (figs. 5.1 and 5.2). Although the GUI did not become the dominant personal computer interface technology until Windows became stable and popular in the business world with the release of Windows 3.1 in 1991, the GUI was anticipated as the inevitable future standard from the mid-1980s on.

For most computer users, the GUI was a vast improvement in terms of usability and functionality over text-based systems. Concurrently running programs (software multitasking), standard menus, detailed graphics, and intuitive computer control via the mouse and desktop metaphor made computers more user-friendly. The idea behind an interface like the GUI comes out of early computer developers' belief that the computer had the potential to be a convivial technology and a technology of intellectual augmentation. Its development at Xerox PARC by computer researchers such as Alan Kay was motivated by desires to make the computer usable for everyone. Kay wanted to build a computer interface that could be used by

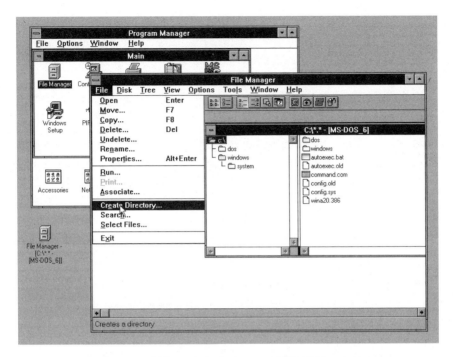

Figure 5.2. Creating a new folder in Windows 3.1.1. With GUI operating systems, creating a new folder required the user to click through various icons and menus with the mouse. This operation was intuitive for most users but created an obstacle for screen-reading technology. Courtesy of Matt McGuire.

a child and would be intuitive to even completely new users. In order to realize its potential to augment human ability and improve human lives, the part of the technology that the user interacted with—its face, so to speak—had to be something users would both want to and be able to use. Kay described the central role of the interface: "What is presented to one's senses *is* one's computer."[6] The computer interface is the computer for the average user; it is how they control and navigate the computer and how they experience the information that is outputted. Most users never go behind the scenes to program code, and much of the computer's hardware is black-boxed and hidden away inside the case. The interface is what the user interacts with, and with the GUI and its metaphor of the desktop, the user is further removed from the computer architecture itself and experiences the computer as something familiar and commonplace. Users have less access to what is actually going on within the computer and its code but gain the benefit of increased usability.

The benefits of the GUI as intuitive to operate and recognizable held true for most users with disabilities as well. GUI systems had standardized menus, and the same commands in the same location were common across programs, which made learning how to use the computer easier for everyone. IBM researcher James Thatcher described, in the *Braille Monitor,* how this move toward standardization at IBM would work for different users.

> Basically one uses the same ways of navigating in many different applications. Text-mode programs were heading that way as they added menu bars, pull-down menus, dialogs, and the like. Still navigation in text-mode Word-Perfect 5.1, Lotus 1-2-3, and Quicken were all different. The GUI versions (OS/2 and Windows) of these applications do in fact have a common interface. The ways to get to menus, to move around menus, to pull down menus, to interact with dialogs are all the same.[7]

The standardization in GUI software controls made learning how to use the personal computer easier for everyone. In particular, the increased ease of use helped people with learning disabilities who struggled with text-based computers. For people with disabilities that required them to use adaptive devices to control the computer, standardization provided program controls that adaptive device manufacturers could expect to be the same regardless of the software being used. Being able to anticipate standard menus and controls made it easier for developers to make adaptive devices that would work with different programs, and it allowed users to make assumptions about how software operated, even with programs the device or user had never encountered before.

The GUI was a particular benefit for those users for whom graphics were more usable than text, such as those with certain learning disabilities or those who found pointing and clicking with a mouse easier than typing on a keyboard. In order to understand the user's perspective on why the GUI could be such an improvement, I examine the account of Mike Matvy, a psychologist with learning disabilities that affected his ability to read and write. He provided a detailed report on his experiences with personal computers to the Alliance for Technology Access for their technology and disability symposium in August 1990.[8] After acquiring a job where he no longer had secretarial support that allowed him to dictate written materials that someone else would type up, Matvy approached an ATA resource center to learn how to use a computer that would read aloud printed materials and help him take notes and organize records. He learned how to use both an IBM computer with MS-DOS and an Apple Macintosh. Matvy had difficulties with the IBM computer's text-based interface: "When I started on the IBM I found that reading and spelling was required every step of the way. As soon as I turned it on I have to start sounding out words and gessing at spelling."[9] Furthermore, he said, "It seam odd to me also that I can learn how MS-DOS works and how to use it to talor specific setts of commands (macros) to do clever things, yet I can not remember the simple letters and sintax required to put MS-DOS to use."[10] The interface's reliance on text, which continuously needed to be read and entered in order to operate the DOS computer, acted as a barrier. As long as it used a text-based interface, the personal computer remained an inaccessible technology for him.

The Macintosh, however, utilized a GUI with symbols such as desktop icons. Matvy learned how to use the Macintosh quickly and even found that the text present in its menus was easy for him to memorize, since it was standardized across applications. He described his success using the Macintosh: "I could also see why I was able to move through the MAC system with such speed and eaze. It is built on a visual system, but it requires no spelling and verry little reading to oparate it. The fue writen words in the pull down minues and the dialog boxes are repeated identicly in all aplications. They are also kept with in a pictoral context which helps me know what the words are."[11] For someone with learning disabilities that made working with visual text difficult, the Macintosh's GUI provided a more user-friendly and intuitive experience. The interface itself functioned here as an accessible technology, allowing Matvy to operate a personal computer in a way that worked with his abilities. Whereas the GUI was an improvement and convenience for most users, for Matvy it was a necessity.

Although people with certain kinds of disabilities eagerly anticipated the positive aspects of GUI technology, blind and vision impaired users looked to the future of the GUI with trepidation. Built into the concept of the graphical interface was the assumption that users could see the screen. This was a mass market, general technology created for a sighted user. In order to accommodate the needs of people with vision impairments, third-party assistive technologies had to change the way the computer output was experienced, namely, by using screen-reading software that allowed users to hear the displayed information.[12] Blind users' worries about the GUI dealt with the way screen-reading technology of the 1980s functioned and its perceived limitations. As explained by James Thatcher, of IBM Research (the company's global network of science and technology research labs), screen readers already had problems when they encountered graphics in a text-based operating system. They were unable to translate the graphics into useful information, so they would skip over graphical modes of software or be blocked entirely from certain applications. For example, up until the early 1990s, blind users were unable to access Flight Simulator entirely, as well as features in Word-Perfect and Lotus 1-2-3 when those programs entered a graphics mode.[13]

Problems translating the graphics of a GUI were not the only worry, however. Doubts also existed about the capability of screen readers to understand how text is displayed on a GUI, which uses pixels instead of ASCII characters. With a text-based interface, a screen reader accessed the display memory and translated the ASCII codes stored there into spoken words. With a GUI, however, the display memory only contains information on the status of each pixel, without any information on the content of what is being displayed. GUI screen readers required a way to work with this output information differently than how they had operated previously. In 1989, Herb Brody described the reliance on text-based systems for screen readers in an article in *PC/Computing* magazine: "In fact, virtually all PC adaptive equipment for the blind operates in the character-based DOS environment."[14] Any graphics the screen reader encountered on an IBM-compatible computer would be ignored. With a Macintosh, however, screen readers were, at this time, unable to translate the pixels on the display into text. Brody described the fears about how the coming change from text to GUIs would affect blind users: "The day is approaching when graphics cannot be ignored. The PC industry's move to graphical user interfaces is arousing concern among the visually disabled—and with good reason. . . . The more graphical the interface, the less translatable it is into speech." Brody's predictions of the coming importance of graphics in computing were correct, as were his worries that the screen reading technology he knew would be unable to handle graphical interfaces. The solution for blind computer users would eventually be a technological one, accomplished during the slow transition from text to graphics.

By the early 1990s, users looking at personal computers with either text-based interfaces or GUIs could still select whatever worked best for them individually, as a number of options existed. While Windows 3.1 (released in 1992) was becoming the dominant personal computer operating system, MS-DOS was still being developed and sold. Windows ran on top of MS-DOS, which allowed DOS applications to continue to be run on a Windows computer (including programs and devices for people with disabilities, such as screen readers). For blind computer users, fears of never being able to access GUI computers finally proved false. In late 1989, Berkeley Systems released the first screen reader that worked with a GUI, OutSpoken, for the Macintosh. IBM released Screen Reader/2 for its OS/2 GUI operating system in 1991. One of the first GUI screen readers, it was developed in communication with blind IBM employees and the National Federation of the Blind. Although it was an important technology for blind users, Screen Reader/2 would never reach a large market, as the OS/2 operating system never captured a mass consumer base, and a few years later Windows came to dominate the personal computer market. The first screen reader for Windows, SlimWare Window Bridge from Syntha-Voice Computers, was released in 1992 for Windows 3.1. A few years later, Syntha-Voice released the first screen reader for Windows 95.

The solution to the problem of how to enable screen readers to translate information on a GUI system came from rethinking how the screen reader accessed information. In order for a screen reader to translate GUI information into text, Berkeley Systems developed what it called the Off-Screen Model for the OutSpoken screen reader. Instead of using the information contained in display memory, which worked for text-based interfaces, the Off-Screen Model allowed information to be intercepted before it went to the display and stored separately in memory for the screen reader to access. As opposed to reading what was being displayed on the screen after the fact and thus being unable to translate pixels into text, the screen reader used this separate memory created by the Off-Screen Model to read what was being sent to the display before it was turned into pixels.[15] This innovation allowed screen readers to function with a GUI for the first time, and although other obstacles relating to the shift from text to graphics remained, blind computer users were no longer faced with what had seemed an insurmountable barrier to using computers with the new interface.

The OutSpoken screen reader not only offered a way to translate GUI text to speech but also provided some of the same access to controlling the Macintosh operating system that a sighted user would have, thereby preserving the spatial layout and navigation that a GUI offers. OutSpoken translated names of icons and simple graphics. In order to translate graphics, the screen reader needed access to information containing some sort of label of the graphic's content. This could be the name of an icon or a nonvisible label that the operating system or software made available. To navigate the computer, users controlled the cursor with the number pad on the keyboard, instead of with a mouse. The user pressed a function key, and OutSpoken would speak the location of the cursor on the screen and indicate when the user reached the edge of the screen. However, this first GUI screen reader was not perfectly integrated into the Macintosh operating system. It did not work with all software applications, did not offer the option to use the mouse, could not translate complicated graphics, could not be adapted to any specialized needs of the user, and did not work with other operating systems. The TRACE Center praised

OutSpoken for being the only screen reader at the time to work with a GUI system but criticized it for not providing vision-impaired users with the full benefits of the GUI, as it only used speech to communicate information. The TRACE Center hoped for technology to take further advantage of what the GUI offered by utilizing other sensory information, such as tactile output or locational sounds, to allow the vision-impaired user to navigate the computer using representational output in the same way a sighted user used representational graphics.[16] Although such alternative sensory technologies were possible, their development had met with limited success, and screen readers remained the dominant accessibility technology for people with vision impairments.

For a few years during the early 1990s, these multiple options of types of computers and interfaces existed, and people with disabilities could figure what best fit their own individual requirements. A 1993 booklet published by the DCCG, "Access to Computer-based Telecommunications for People with Disabilities" included recommendations for users choosing between IBM-compatible computers with DOS, IBM-compatible computers with Windows, or Apple's Macintosh computers: "For people whose physical disability precludes use of a mouse or trackball, a DOS system might be preferred because it does not require either one. . . . However, people who are unable to enter text from a keyboard, but are able to use a cursor with a mouse, joystick, or trackball, may prefer a Macintosh or a Windows interface, which can be faster and simpler to use."[17] The DCCG's advice here was similar to the advice they had been offering since the early 1980s: people with disabilities needed to find what worked best for them individually. For people with vision impairments, the DCCG still suggested a text-based system, although screen readers were beginning to be developed for GUIs at this time: "There is a wide range of screen readers for DOS systems, and currently three choices that will read the graphics-based systems. Given the choice, a DOS system presents fewer hurdles for people with visual impairments."[18] Accessible technology was beginning to catch up with the change in interfaces, but in terms of ease of use, during the early 1990s, a DOS or other text-based system still worked better with available screen readers than a GUI operating system.

By the mid-1990s, the GUI became the dominant personal computer interface, although it still continued to present problems for vision-impaired users. Users addressed the problems directly with operating system manufacturers. In particular, a dialogue took place between Microsoft and organizations for people with vision impairments concerning Microsoft's failure to respond to the need for fully functional screen readers for Windows. The first screen reader for Windows was not available until 1992 and was developed with little help from Microsoft. In 1994, the head of Microsoft's Accessibility and Disabilities Group, Greg Lowney, responded to these criticisms in a letter to the National Federation of the Blind, admitting Microsoft's culpability: "Windows has probably done more than anything else to earn Microsoft the enmity of the blind community. Microsoft has been both hated and feared by many people because we were promoting a graphical operating system without making sure that it could be used by people who are blind, and the results have been disastrous for many people."[19] Although Microsoft had not built in adequate accessibility or easy ways for third-party developers to create assistive devices for Windows, Lowney argued that, in theory, Windows should eventually be more accessible than MS-DOS because the standardization of a GUI should allow screen readers to work across all software applications. Microsoft's growing operating system monopoly played a dual role for users with disabilities; it aided the spread of standardization and its benefits, but it also compounded accessibility problems when they arose, as users had fewer alternatives. To assist its stated goal of improving accessibility, Microsoft began to release documentation to third-party developers and to respond to users' feedback about Windows.

A few years later, after the successful release of Windows 95, screen-reading technology had improved to the point where it had nearly full access to the GUI operating system and popular software applications. A 1997 article by Kenneth Frasse in *Access Review* compared the capabilities of available screen readers with Windows 95 functions and control of popular software applications (Microsoft Word, Microsoft Excel, Internet Explorer, and Netscape Navigator).[20] The tests consisted of everyday tasks of multiple steps that users would want to perform for business, home management, and recreation purposes. These tasks included installing the screen reader from DOS, navigating across the desktop, copying and creating shortcuts, setting the date and time, changing the display appearance, copying files, formatting a floppy disk, opening files in Word or Excel and reading through them, running spellcheck, and navigating to common Web sites and reading their contents. The review rated the screen readers on whether they allowed the user to fully complete the task, complete it with varying severity of problems, or attempt the task but be unable to complete it. It also considered whether the screen readers lacked the functionality to even start the task. Only one of the screen readers performed well. GW Micro's Window Eyes was successful in almost all the tests, and it was found to be capable of navigating the GUI operating system and the software applications. Window Eyes and its chief competitor, JAWS from Freedom Scientific, still dominate the Windows screen-reader market today.

During the history of its development, the graphical user interface presented new opportunities for some users and challenges for others. The GUI created new forms of access to personal computers for some people, such as those with learning disabilities who previously had struggled with text-based interfaces. It also provided new ways to organize information and browse, using symbolic, graphical representations that made personal computers more user-friendly and intuitive for most users. However, this intuitiveness was based on the GUI's use of visuals and, with it, the assumption that users could see the computer screen to use it. The GUI created new barriers for users with visual impairments, and its gradual development allowed worries to grow and innovative technological solutions to be found. By rethinking how screen readers accessed output information, GUIs were made at least somewhat accessible. The standardization and operating system control that came with the GUI also made screen readers capable of, in theory, functioning across software applications by allowing expectations of consistent navigation and control. Blind users were blocked from fully enjoying the advantages of the GUI, however, during much of the course of its development, mostly as a result of lack of attention from operating system developers. Problems for screen readers continued with the 1997 public accessibility disaster of Microsoft's Internet Explorer 4 (which I discuss at the end of this chapter). Even once GUIs were made to work with screen-reader technology, individual software applications continued to cause problems by not following accessibility standards. However, not all major software companies resisted making their applications accessible to people with disabilities or considered it an afterthought. Turning from operating systems and interfaces to specific software applications, I examine the work done by one successful software developer to improve its own accessibility through a partnership with the Alliance for Technology Access.

The Alliance for Technology Access and Brøderbund Software

Brøderbund Software was one of the longest lasting and most commercially successful software companies of the 1980s and 1990s. Douglas Carlston, a lawyer turned software programmer, founded Brøderbund as a video-game producer and publisher in 1980; it quickly grew to become a dominant player in software publishing for computers and game consoles.[21] Developers at the company also concerned themselves with issues of accessibility as a way to make their products work with as many users as possible.

Staying strong throughout the 1980s, Brøderbund's sales then increased dramatically starting in the early 1990s as the invention of the CD-ROM allowed for the distribution of far larger and more complex games and other kinds of software. In 1992, Brøderbund released "Just Grandma and Me," the first in its interactive, animated Living Books series and an international bestseller.[22] A year later, Brøderbund published *Myst,* which became one of the best-selling personal computer games ever.[23] Brøderbund's success allowed it to operate a creatively free environment for its programmers, permitting the company to push the boundaries of computer programming.[24] Brøderbund marketed to a mass consumer base, publishing software for entertainment, education, and home management. Yet it did not develop computers itself; Brøderbund had to fit its products onto systems created by other companies. As it sought to attract more customers and improve the usability of its technology, the company began to pay specific attention to the needs of computer users with disabilities.

At the time of its peak commercial success, in 1995, Brøderbund turned its resources and innovative energies toward improving accessibility for people with disabilities, starting a partnership with the Alliance for Technology Access and becoming a vendor member of the Alliance. The goal behind this partnership was to promote a universal design approach that would make software applications work for as many people as possible. The ATA provided a definition in 1996 of its understanding of what universal design should entail and what it encouraged software companies to implement: "(1) that all products are robust and flexible out-of-the-box, with built-in access features; and (2) until off-the-shelf full access is a reality, that products work smoothly with third-party assistive technology such as touch screens, alternative keyboards, screen readers or voice input systems."[25] Enacting principles of universal design marked a move away from any notion of the universal human as the intended user; instead of generalizing all users into one, universal design calls for a pluralization of imagined users, encompassing all individual needs. Universality is achieved by accommodating all possible differences among people.

The Alliance observed software developers increasingly adopting ideals of universal design and sought to promote the trend. More software applications were becoming available that worked for more users and responded to individual needs by providing options. In its book, the ATA laid out its view: "The worlds of assistive and conventional technology are blending, and a new generation of products is emerging—products designed to be used by *all* people. A number of companies are aware of the need and are designing products with universal access in mind. One company in particular, Brøderbund Software, recognizes the need for designs that provide the greatest function for the greatest number of users."[26] The ATA used the phrase "universal access" as a way of focusing specifically on including people with disabilities as intended users. As with "universal design," however, nondisabled users are also implicitly included. The binary classification of disabled and nondisabled users is replaced by a set consisting of the diverse needs of all users.

As accessibility features became more commonplace and standard in personal computer technologies, the need for specialized devices and software diminished. Although universal design could never function absolutely—there is no one input device that all people can use, no matter how accommodating it might be—even small features of flexibility and options built into computer technology made it more usable generally. By building a philosophy of universal design into the initial programming process and working to understand the diverse needs of their users, developers would have a larger audience for their products. Brøderbund recognized the benefits of using universal design and contacted the ATA to have its own products tested and evaluated for their accessibility.

The ATA suggested two different approaches to using universal design to make software accessible for people with disabilities. First, allow external hardware input devices to work with the software, and, second, provide internal flexible options for multisensory or expanded sensory output. With the first approach, software developers could ensure that their applications would work with assistive devices such as alternative input devices: "Making software more accessible can mean making sure that programs will work with specifically designed products that enable individuals to interact with software by using alternatives to the standard keyboard, mouse or method of display. Some examples of these include products that read aloud text on the screen, that allow a user to control the cursor by moving their head or raising an eyebrow, and that let someone enter text by clicking a switch or speaking directly to the computer."[27] The first, immediate step of accessibility is physical access to controlling the computer. As the 1984 White House task force had recommended, all computers needed access points available so that users could plug in and use whatever devices they needed to operate the computer.

Once the user had access to the computer itself and the ability to physically control it, software developers could then build alternatives into their programs which would give users flexibility in terms of how they could understand the program's output: "Within the software design process, making software more accessible means ensuring options are built in that offer alternative ways to work with a program, such as text options to augment dialogue, the ability for any program to read aloud text on the screen and the ability to enlarge the standard size of print and graphics."[28] One example of such an option was one of the first accessibility features that Alan Brightman and the Office of Special Education and Rehabilitation were able to have Apple engineers build into the Macintosh—a visual error indication instead of only an auditory one. This understanding of different needs was missing initially with the transition from text interfaces to graphics. Although developers sought to improve usability for everyone, they built in sensory assumptions of how the technology would be used and neglected to include options for users who could not see.

Giving the user choices in how to receive computer output creates access for people with certain disabilities where previously there might have been an insurmountable barrier, and it also increases usability for all users. Someone who could see the screen but whose eyes tired trying to read small text could have ways to enlarge it, or someone using a computer in a quiet environment could turn off spoken dialogue and read subtitles instead. Brøderbund explicitly noted this advantage of universal design in creating products in the announcement of its partnership with the ATA: " 'We have consistently found that designing software that's more accessible to those with disabilities also makes the software more intuitive and easy to use for the balance of the market,' said Bill McDonagh, President and Chief Operating Officer of Brøderbund. 'Besides being the right thing to do, it makes all of our products just a little better.' "[29] By focusing on universal design as a philosophy to be built into the programming process, Brøderbund hoped to increase its customer base and improve the users' experiences with its products. Changing its programming methodology was not the only way that Brøderbund sought to improve software accessibility, however.

The concept of access has multiple meanings and ways of accomplishing it. Beyond building accessibility features into its products, Brøderbund also worked with the ATA to entrench the idea of accessibility into the computer industry. Their partnership sought to influence accessibility in other companies and to provide communication tools that would allow developers to better incorporate accessibility features. Being a major player in the software industry, Brøderbund hoped to encourage other companies to follow in its footsteps and offered some of the tools to help them do so: "Brøderbund and the Alliance are currently working to incorporate more access features into future Brøderbund products and hope that this announcement will encourage other software developers to make a similar commitment to creating products with inherent accessibility features. Brøderbund has already prepared preliminary guidelines for accessible software design and is applying these to current products in development. Both organizations are prepared to provide assistance to other software developers that are interested in developing software that includes this important market."[30]

Brøderbund's long-term success in the volatile software industry and reputation for being a creative haven for programmers who produced innovative software put the company at the forefront of the industry, a place from which it hoped to guide other software developers. Brøderbund wanted to use this leverage to spread the value of universal design and accessibility for people with disabilities. The company also instituted corporate-wide structures to make including accessibility features part of their standard development process and to communicate that work internally and to other parts of the software industry: "They have established an internal bulletin board to which information related to access is posted for everyone's use. And engineers, product managers, and sales staff are collaborating with other vendors in the field to create solutions that mean greater access to Brøderbund software."[31] While it is unclear to what extent it was able to impact other software developers in improving their accessibility, Brøderbund did demonstrate how it had taken these goals of accessibility to heart.

One of the outcomes of Brøderbund's partnership with the ATA was a 1997 evaluation of the accessibility of its products.[32] The tests went beyond just determining if a user could operate certain aspects of the software, instead focusing on the skills the user would gain by performing certain activities with it. Specifically, the evaluation looked at process and academic skills that children with learning disabilities needed to develop. The key question in evaluating the software was, "Does this activity provide the opportunity to foster these skills or qualities without facilitation or intervention?"[33] The emphasis was on the skills the children ought to be gaining and their ability to operate the software independently. The skills tested for included general "process skills," such as concentration, hand-eye coordination, auditory perception and discrimination, visual perception and discrimination, memory, communication, logic, and strategic and creative thinking. The academic skills included problem solving, reading, writing, spelling, and mathematics. In addition, the software was tested for the presence of common accessibility features and its compatibility with assistive devices. The ATA conducted the software tests using an Education Advisory Team made up of members of Alliance centers who had expertise in learning disabilities, working with children, and product testing. They extensively evaluated five of Brøderbund's educational software titles that targeted writing, reading, math, logic, and drawing.

The ATA's Education Advisory Team published the results of their evaluation of Brøderbund's software in the form of a poster for educators and parents to use when selecting appropriate software.[34] A chart detailed how each of the five software applications promoted the improvement of the process and academic skills, and it also listed the common accessibility features found in each program. In addition, the poster provided information on how well titles in the larger Brøderbund catalog worked with assistive devices. The assistive device tests were conducted on Macintosh and Windows systems, along with some testing for DOS versions of software. These tests included titles from Brøderbund's educational software, the popular video game *Where in the World Is Carmen Sandiego?,* and the home printing and graphics program *The Print Shop.* The tests determined that all of the software titles were compatible with the tested alternative keyboards, switch input devices, screen enlargement software, touchscreens, and electronic pointing device.

Brøderbund was by far the most prominent software company to become a vendor member of the ATA and work with the Alliance to improve the accessibility of personal computer software for users with disabilities, but they were not the only company to do so. The ATA singled out and praised two general educational software publishers, Edmark and Hartley Courseware, for joining in Brøderbund's efforts to partner with assistive device developers to create custom overlays for alternative keyboards that would allow users easier operation of their programs.[35] Education software publishers tapped an obvious market in adapting their products to include special education focuses. Being able to market the accessibility of their products provided positive press for software publishers and expanded their consumer base to include more people with disabilities.

Alliance for Technology Access Publications

In addition to its work with software companies to improve accessibility, the Alliance for Technology Access also produced two publications during the mid-1990s that codified its views on computers and people with disabilities. The ATA published the first edition of its book, *Computer Resources for People with Disabilities: A Guide to Exploring Today's Assistive Technology,* in 1994, providing a guide for people looking for accessible personal computer technology and coalescing many of the ideas and values that had driven the ATA since its founding. The book immediately met with success; within eight months after its initial publication, it was already on its third printing.[36] Two years later, the group published a revised and updated second edition, with a third edition following in 2000 and a fourth in 2004.

Like similar books for people with disabilities, *Computer Resources* does offer information on useful resources, such as companies that produce certain kinds of technologies, but its main focus is on how people can approach computer technology as a means to solve specific problems. It includes case studies of real-life users and their experiences. Employing a celebratory rhetoric centered on individual abilities, solutions, and raising expectations, the book addresses the computer user with disabilities directly. The ATA's perspective emphasizes the role of technology as empowering people to achieve their individual goals and celebrates the progressive trend toward greater accessibility and more usable technology. People with disabilities are the book's intended audience, and in it the ATA pushes for users themselves to make decisions, where possible, about the technology they will use.[37]

The ATA believes that the expectations that these users have for themselves and their use of technology are essential to making technology able to change people's lives for the better: "The success of technology has more to do with people than machines. All the right parts and pieces together won't work miracles by themselves. It is people who make technology powerful by creatively using it to fulfill their dreams. The evolution of the field of assistive technology is more about the evolution of people and their expectations than it is about circuitry." From the ATA's perspective, people with disabilities have had low expectations placed upon them for too long; in order to fulfill their own personal goals in life, people need to throw off what is expected of them and embrace seemingly "unrealistic" expectations. Historically, the Alliance sees an evolution in the expectations people with disabilities have had placed on them and have placed upon themselves. "There has never been a better time for an individual with a disability to challenge all of the stereotypes and notions of 'unrealistic' expectations existing in our culture. Not only do we have the right to envision and develop unrealistic expectations, but we have the right to achieve them." Technology could provide the opportunities for people with disabilities to live up to expectations, deemed unrealistic for them, that were utterly normal for nondisabled people. Assistive technologies marked a step toward equity, in making the expectations for people with disabilities more like those for people without. Expectations and technology fed into each other here; the expectation that technology would be there to help people accomplish their goals furthered the development of technological improvements as developers' awareness of the needs of people with disabilities increased and more people with disabilities entered careers in technological development.[38]

However, the ATA did not intend to send a message that technology, especially computer technology, would always be a quick or easy fix. The book emphasizes the individual nature of computer technology; what works for one person and their circumstances may not work for another: "You may be able to reduce confusion by realizing that technology decisions are very personal. There is no one best computer, no one best software title, no single universal access device. There are only tools to be found that work well for you in your circumstances."[39] Universal design is an ideal to be encouraged, but it is not actually capable of achieving the results it strives for. No technology can be designed to meet the needs of all users. What technology allows is a way to empower people to achieve their goals. The Alliance treated technology as a neutral tool whose use is determined by the creativity of people. The values embedded in a technology, in its design, marketing, and distribution, are absent from this account. Consistently optimistic, the book neglects to mention some of the negative aspects of technology, only showing users ways that they can improve their own lives.

As a way of focusing more on empowerment than obstacles, the book asks users questions about what they are able to do (their abilities) and what they want to accomplish with computer technology, as well as what difficulties they might have. Unlike the books by Peter A. McWilliams and Frank G. Bowe, the ATA book is not organized by type of singular disability but by specific questions regarding degree of different abilities. For example, instead of listing technologies appropriate to blind or visually impaired people, the ATA provides charts that suggest different tools for people depending on the degree of sight they have. This organizational structure allows users to pick and choose tools to solve different problems they encounter when using a computer, instead of focusing on disability categories, an approach in line with the universal design ideals that had been growing since the 1980s.

In this book, the ATA shows the special place technology has in the lives of people with disabilities, as it not only allows them to overcome social barriers preventing equal participation but also can provide opportunities they might not otherwise have available. The ATA describes this in terms of people being able to make choices about how to live: "In every sense of the word, *empowerment* is an attitude available to everyone with a disability. The law provides the legal rights and sanctions, but technology and imagination provide the real capacity and ability to choose, to act and to invent your future."[40] Legislated antidiscrimination was insufficient to allowing people with disabilities equal participation in society; technology could provide the next step in fulfilling the promise of civil rights legislation such as the ADA. The ATA also emphasizes the role of individual imagination here; it is a personal tool people can use to solve technological problems in order to reach their goals, but the necessity of imagination also indicates an awareness of the difficulties in making these technologies work (that they would require imagination in the first place).

To stress the importance of individual choices and experiences in using technology, the Alliance's book includes both general advice and specific case studies of real people. The accounts of actual computer users with disabilities give a concrete human perspective to this guide. Notably, Stephen Hawking wrote the foreword, in which he discusses the role technology played in allowing him to communicate with other people: "This book is about problems of expression and communication, and how to solve them. . . . I, and thousands like me, have been helped to communicate by modern technology. Indeed, the fact that I have been asked to write this foreword is a sign of what technology can do."[41] Hawking described, in detail, the problems inherent in trying to communicate without speech, talking about it in terms of rate of information flow. Speech can produce between 120 to 180 words per minute, whereas average typing is limited to between 40 to 60 words per minute. His ability to communicate was further complicated by his disabilities; he could only operate head or hand switches, not a full keyboard, which made spelling out each word a time-consuming process. Computer technology, however, allowed Hawking to select whole words or phrases, instead of individual characters, at a rate of about fifteen words per minute. Though this was not as fast as he desired, he remained optimistic: "the promise of computer technology is that improvements are always in development."[42] One improvement he observed was in speech synthesizers. Hawking asserted that everyone wants to sound human, not like a machine or cartoon, and that speech synthesizer technology was finally at the point where that was becoming possible.

Hawking used this example of his search for technology that allowed him to speak to demonstrate what the ATA's book attempts to provide for others: "I hope others find in this book the inspiration and the technology, hardware and software, that can help them to communicate better—to express their human-ness."[43] The need to express one's humanness was fulfilled for people such as Hawking through the use of communication technologies, which made their speech intelligible and made them more able to participate in society. Equating humanness with verbal communication capabilities, however, created a singular definition of what it means to be human, while instantiating the necessity of using assistive technologies that provide for certain methods of communication. In the book, the ATA echoed the importance of people being able to communicate as one of the actions made possible with accessible computer technology. Their rhetoric celebrates computer technology's furnishing of abilities an individual might otherwise lack, without acknowledging what it might mean for someone to be dependent on technology in order to perform such expected aspects of human socializing. After Hawking's account, the rest of the users featured in the book are actual clients or members of ATA centers who present shared stories of finding individual solutions to problems through the use of computer technology.

In one account, Tom, a former rehabilitation professional and cofounder of a self-advocacy group for people with disabilities, relates how using technology helps to erase the differences that make it hard to participate in society and to embrace those that make people unique.

> What I am and who I am comes from my interaction with the environment. This technology—in spite of some of the problems with it—is enabling. I see young kids with cerebral palsy and speech difficulties in school using speech output. I see technology helping me to be seen as an equal. I am deeply impressed by how much more equal people with disabilities are because of technology. They have the power to communicate and the power to be more accepted and acceptable. People are less different. Technology allows people to be different on their own terms, rather than on society's terms.[44]

For Tom, as for Hawking, communication is how people express that which makes them human, and technology is the means through which such communication becomes possible. The ATA reiterates its point about the necessity of communication: "Everyone needs to connect with other human beings, especially youngsters with disabilities. Technology has the power to bring people together by providing them with the ability to interact and communicate in new ways."[45] Online services, which were growing increasingly available at the time of the book's first publication, provided one of the new forms of communication that computer technology was making possible, and communicating online was another way that computers helped people like Tom to feel like they could be seen as equal. These new technologies also broadened the types of communication that were possible with the personal computer, allowing for some of the erasure of the stigma of disability that Murray Turoff had seen as a benefit of computerized conferencing.

Underlying these concerns about being able to interact with other people regardless of disability was the desire on the part of people with disabilities to be seen and treated the same as everyone else. That people with disabilities were not inherently different was also at the core of universal design. People need or want to use technology differently, and these different ways can be accommodated by building in flexibility and options for how a technology can be used. The differences between people matter for universal design insofar as developers need to be aware of the different abilities people possess, in order to understand their individual needs. There is no hard distinction between people with disabilities and those without; universal design attempts to accommodate everyone's needs, no matter what those might be. The ATA emphasizes this point in its book by making it clear that the group's efforts are focused not only on specialized, assistive technologies but also more generally on methods of creating access to conventional technologies.

> Much of what individuals with disabilities want is simply access to conventional technologies. You want to be able to write with a word processor. You want teachers to be able to read your writing. You want to publish a newsletter. You want your child to be able to draw and experience the process of creating pictures. You want to be able to create and perform music on a synthesizer. You want to play the latest computer games. You want to work a cash register. Your daughter needs access to a patient tutor for learning her multiplication facts.[46]

People with disabilities and those without share many of the same everyday goals that are accomplished with technology. Universal design offers the means through which technologies can be made to work for people regardless of disability. In its book, the ATA argued that the growing acceptance of universal design was beginning to change the way computer technology was developed, in terms of usability: "The distinction between assistive and conventional technologies is becoming less clear as the concept of universal design is incorporated into conventional technology. Both fields are broadening and converging. What is a necessity for some is convenience for all."[47] As developers became more aware of the needs of users with different abilities and the benefits that incorporating universal design into their development process could have, the Alliance saw real improvement in the flexibility of computer technology.

After the success of their book publication, the ATA released a short video, *Quality of Life: Alliance for Technology Access* (1995), promoting the work done by Alliance centers and drawing attention to the need for accessible technology in the lives of people with disabilities. The cost of producing the video was underwritten by IBM's Special Needs Systems with help from past ATA supporters Brøderbund Software, Pacific Bell's Deaf and Disabled Group, IntelliTools, and Living Books. One of the major themes presented in the video is individual empowerment for people with disabilities through control over one's own life with the aid of technology. In the video, Jackie Brand explains the value of independence within the ATA, "At the heart and the soul of the Alliance for Technology is a belief in certain values that say it's time for people with disabilities to take their place in the world alongside everybody else, to determine their own futures, to make decisions about their own life and to have the chance to achieve their most outrageous dreams."[48] Brand saw the ATA as providing places where people could find technology that will enable their independence.

Jean Issacs, the educational director of the Alliance center in Lexington, Kentucky, echoed the importance of technology allowing people to take control of their own lives, particularly for children with disabilities who might otherwise have few opportunities to make decisions for themselves. Issacs explained that being able to watch children have such an experience is one of the reasons she works for the center: "You know, it's that special moment that happens when we get a kid in here who's never experienced a switch toy and for the first time controlling their environment when they hit that switch and the clown laughs or the train shouts 'I think I can, I think I can.'"[49] Issacs's account was similar to the way Brand talked about her daughter Shoshanna's early experiences with computer technology. The Brands first taught Shoshanna to use a Unicorn keyboard by programming it to respond to any press of it, thus showing their daughter that it was her decision to act that the computer was responding to and that she could have an effect on her environment. The versatility and adaptability of computer technology allows it to be responsive in this way; a single switch operated by any muscle in the body can enable a user full control of a personal computer to do with it as he or she wishes. For people with disabilities who may otherwise struggle with control over their lives—either because they are assumed to be unable to do things or they face social barriers that remove their choices—the independence provided by the computer offers one place where they can be fully in control.

The *Quality of Life* video also emphasized another important aspect of the ATA's mission, that of bringing people and knowledge together through the network of ATA centers. Christy Blakely, client of the center in Littleton, Colorado, and parent of a child with a disability, talks in the video about how she has been able to share experiences with others in order to not have to solve every problem on her own: "The Alliance gave us people and a place to seek out the answers for what was available, for when we came up with a problem. It also allowed us networking capabilities with other families, other parents, other families so that we weren't recreating the wheel. So that we were saying 'Has anybody ever come up against this?' and they could direct us to another family that maybe had done that or who had experienced what we were going through."[50] Much of the history of people with disabilities in American society is one of isolation. By offering a place to find commonality with others, the Alliance helped to fight against one of the stigmas of disability and provided a way for people to solve their problems using both technology and each other.

The ATA believed that technology was the means to provide access to participation in society for people with disabilities. Blakely praises the role technology played in opening up the world for her daughter: "It really allows them access to the world and it's going to make a population of kids that have a chance to be productive adults."[51] Technological accommodations had helped people with disabilities to find employment and participate in social activities since legislation in the 1970s required them. For parents like Blakely, technology was continuing to offer children with disabilities new ways to engage with others, learn, and look to a future of greater independence and social involvement.

From the perspective of a computer user with disabilities, Dr. David B. Rogers discusses in the video the role technology plays in allowing him to continue as a clinical geneticist after becoming disabled: "Well, I'm using technology to access the world. Whether I'm working or playing, writing a letter or playing a game. Or doing a literature search like I'm doing that right now."[52] His positive experiences with the ATA led Rogers to becoming a board member at the ATA Computer Access Center in Santa Monica. As with the ATA's book, this video spread the message that people with disabilities were capable of living productive, independent lives and that computer technology was the tool that could enable them to do so. The Alliance's overall success and growth during the 1990s was matched by similar success and growth at its individual centers, particularly the Disabled Children's Computer Group.

The Center for Accessible Technology

The 1990s saw the Disabled Children's Computer Group expand, becoming inclusive of more computer users and utilizing new kinds of computer technology to benefit them, while continuing to remain locally focused on the Berkeley area. Given that the population they served included adults with disabilities (including some of the original clients who were now grown up) just as much as children, the board proposed a name change and broadening of the mission and goals. At a January 1992 meeting, it was proposed that the group keep the D.C.C.G. acronym but alter its meaning to include adults.[53] One board member wrote down possible ideas, among them to change "children's" to "community" or "citizens." The group members sought continuity with their past and the community they had established, yet they needed to become more explicitly inclusive. In 1994, the DCCG adopted an entirely new name, the Center for Accessible Technology (CforAT). This new moniker not only better reflected the group's focus on people with disabilities of all ages, but it also referred to the work the group did with all kinds of technologies, not just computers.

With its change in name, the CforAT officially brought adults with disabilities under the scope of its services. In 1995, the group began conducting evaluations and training for adults who were Department of Rehabilitation clients or beneficiaries of worker's compensation. In addition, corporations that dealt with accessible technology could also access CforAT services. These services still took place within the CforAT's resource center, which continued to be where the group connected with the local community by offering a place for people to experiment with technology that might help them.[54] This focus on hands-on technology exploration continued in the CforAT's Open House Resource Sessions, where anyone in the community, as well as technology vendors, could come in and explore technologies and share their expertise with others. In addition, the CforAT had its one-on-one services, now called Guided Explorations, which dealt with specific individuals' problems. The resource center also conducted various seminars and classes for parents, teachers, and children, along with special play groups where children with disabilities could try out accessible toys.[55]

In addition to its locally focused services at the resource center, the CforAT also began larger scale programs in the mid-1990s, such as the PlaneMath project. This project was an example of a type of software-related computer accessibility that increasingly became a priority at this time: Internet Web sites. PlaneMath was a joint effort between NASA and InfoUse to develop an educational Web site in mathematics and aeronautics targeted at children in grades four to seven with physical disabilities. A number of ATA centers worked to help local schools access the Web site, with the CforAT taking the lead in donating staff and other resources. The project attempted to address two needs: to improve math education for students who had disabilities affecting their use of tools commonly found in math classes (such as pencils, calculators, and geometric models) and to encourage children with disabilities to consider careers in aeronautics which required the development of math skills. A Web site was the chosen method to achieve these goals, as argued by the program: "The Internet, with its multimedia and communication capabilities, holds great potential for allowing these issues to be addressed."[56] The site took full advantage of the multimedia technology of the time; activities geared toward teaching students how to use math to solve aeronautics-related problems used graphics, animation, and audio narration. The lessons also taught students information about airplanes and aeronautic careers. The full project only appears to have run for around two years (the final update was in December 1998), but during that time it was recognized by a number of organizations reviewing useful educational Web sites.[57]

The CforAT took part in the PlaneMath project by providing help in constructing the lessons and giving advice on making the project itself accessible to students with disabilities.[58] There were two forms that concerns with access for children with disabilities took here. First, Web sites provided access to new ways to learning, but at the same time they needed to be made accessible in order to work with assistive devices. However, unlike with one's own personal computer, where users can opt between different brands of software to find one that works best for them, they have no control over what the developer of a Web site has chosen to provide. Web sites have to follow accessibility standards in order to work with assistive devices such as screen readers. If they do not, certain Web sites may be inaccessible, without any recourse for the user. (Web sites are less personal than the personal computer as a whole; they are more like public buildings that users visit.) Second, PlaneMath encouraged users with disabilities by explicitly including them; this was done through making the site accessible in a variety of ways and by using people with disabilities as role models in the actual problems that students were asked to solve (fig. 5.3). People with disabilities were subjects in graphics

Let's go fly a kite! Do you know how your kite stays up in the air?

Figure 5.3. PlaneMath problem showing a child with disabilities in an example. Many of the activities on the Web site showed people with disabilities in the graphics, in order to demonstrate their role as intended users. From http://infouse .com/planemath/activities/flykite/kite1.html.

throughout the Web site, demonstrating their assumed place as users of the site and as potential future aeronautics professionals. The participation of people with disabilities is normalized in these math problems, telling the users with disabilities that they can participate in the example activities the same as nondisabled users.

CforAT staff prepared a document that addressed issues of Web site accessibility for the PlaneMath project.[59] It described the built-in accessibility features of PlaneMath, such as large buttons in consistent locations, large text size, clear and relevant graphics, a text-only alternative version of the site to be used with screen readers, and alt tag descriptions of all images. The document also provided general advice on Internet browsers and assistive technologies, as well as technical information on specific assistive technologies that people might want to use with PlaneMath (e.g. text-to-speech software or keyboard access to certain Web browsers). The site as a whole was designed to be accessible with a variety of common assistive input devices, such as alternative keyboards or voice input systems. In terms of output, the only limitation was with the lessons involving animation, as these could not be understood by a screen reader. Text-only pages of the animation narration were provided as an alternative, but the activities that used the animation plug-in were inaccessible to screen readers, and no alternative to them was provided for vision-impaired users.

While it is unclear how many teachers or classes utilized PlaneMath or what kind of impact it had on the education of children with physical disabilities, it does appear to have been well advertised by the organization involved in its development and well received by education-related Web sites. PlaneMath does not list the total number of teachers who registered their classes with the project, but it does list seventeen teachers from all across the country who registered on the site and won randomly chosen prizes donated from corporate sponsors. NASA advertised the project in a 2001 video segment on its Web site geared toward children.[60] The segment demonstrated how to use the PlaneMath site, showing students trying the site out while visiting a museum. InfoUse also presented on PlaneMath at the 1999 CSUN Technology and Persons with Disabilities Conference. Through their involvement with InfoUse and NASA, the PlaneMath project created a venue for the CforAT to reach far beyond the local Berkeley area community and offer resources for teachers and students all across the country. The Internet provided a technological means for the local advocacy group to expand its efforts and write the needs of children with disabilities into the development of educational tools. PlaneMath offers an early example of educational Web site accessibility, before the lawsuits starting in the mid-2000s began to enforce the following of accessibility standards on the Internet.

Brand and Brightman Move On

My history of the development of personal computer accessibility and its relationship to computer users with disabilities ends in the late 1990s, as personal computer technology reached a level of stabilization and accessibility standards became more normalized. This was also the time when two of my main subjects moved on from the organizations they had helped found. Jackie Brand left the Alliance for Technology Access in 1997, and Alan Brightman left Apple Computer's Worldwide Disability Solutions in 1998. Brand had remained as the ATA executive director until this time, when she was replaced by Frederick Fiedler. Fiedler continued in Brand's steps leading the ATA as the parent of a child with a disability, but, unlike Brand, he brought a technical education and management background to the position. Fiedler was a retired Air Force major general, with degrees in engineering and experience managing organizations.[61] His was a far more formal expertise, compared to Jackie and Steve Brands' self-taught tinkering, which had led them to found the CforAT fourteen years previously. However, as with the Brands, it was Fiedler's highly personal connection with disability that brought him to the ATA: "I became aware of ATA when I saw an advertisement to fill this position. Barbara and I are parents of a 27 year old son with a traumatic brain injury and the service ATA centers provide is exactly what the Fiedler family needs. We need to make ATA centers known to more consumers."[62] In addition to carrying on the important work of parents of children with disabilities banding together to help one another and their children with technology, Fiedler also emphasized the role of people with disabilities as both users of technology and as consumers, demonstrating the shift that occurred throughout this history from people with disabilities as receivers of charity to active consumers.

In Jackie Brand's next project, the Universal Service Alliance (USA), an offshoot of the ATA, she used the kind of network building and knowledge exchange she had set up at the ATA to address issues surrounding access to telecommunications technologies for underserved populations in California. Moving beyond serving only people with disabilities, the USA worked to address the growing divide between populations who had access to the Internet and other telecommunications technologies and those who did not.[63] The USA connected a diverse set of activist groups representing Hispanic, African American, Asian American, Native American, and rural communities. Brand viewed people with disabilities as facing some of the same problems as other disadvantaged groups: "And here's a situation where it seemed to me that people with disabilities, though having some unique issues about access, in many ways were exactly in the same shape as many other communities who were also at risk."[64] Her plan for the USA was to have a disability organization take the lead in bringing together people from other populations, utilizing the expertise that disability advocates had developed during the previous two decades promoting accessibility and technology.[65] The assembled network of various organizations held public forums to communicate directly with underserved people about issues surrounding access to telecommunications technologies and worked with telecommunications companies to improve access for everyone.

One of the organization's most prominent actions was to manage the $50 million that the California Public Utilities Commission required Pacific Telesis and SBC Communications to provide for the use of improving access to telecommunications technologies in underserved communities, as a result of the merger of the two companies in 1997. This was a novel approach to handling money that was legally required to be given back to consumers; instead of granting all customers a small amount back on their utility bills, the money would directly address inequalities in access to telecommunications technologies. The Community Technology Foundation of California, created by the USA, managed these funds.[66] Part of the intention was to increase telephone access in low-income and minority communities to 98 percent by providing improved infrastructure, services, equipment, training, technical assistance, and consumer education. Brand hoped the project would encourage these companies' competitors to join them in improving access to telecommunications technologies, a goal backed up by Pacific Telesis's promise to match funds donated by other companies.[67] Brand took the expertise she had developed in disabilities and technology and expanded her focus to help other people who faced disadvantages in accessing and using technology. People with disabilities were not set apart as the only population that encountered barriers to access; other marginalized groups faced similar challenges.

In early 1998, Apple Computer fired Alan Brightman's group, Worldwide Disability Solutions, for cost-cutting reasons as the company attempted to recover from its near bankruptcy two years earlier. Brightman moved on to Yahoo and continues to run its disability group today. In a *Wired* article responding to the firing, Brightman criticized Apple for getting rid of what he saw as a significant benefit to the company: "Apple will say there are no sacred cows, and I buy that. But for what we cost, the return was enormous, not only in reputation and caché, but in dollars. But, I guess it was small enough to leave on the table." Gregg Vanderheiden of the Trace Center echoed Brightman's idea that Apple was losing out on the potential market of people with disabilities by not having an in-house group dedicated to accessibility: "But now that they don't have people dedicated to working on the topic, they won't be at the top anymore."[68] In a more recent interview, Brightman looked back on the Apple firing in more positive terms: "I think at one point we just felt like we were done; that the accessibility was now in the DNA of the company. It didn't need us to grab the lapels of all the engineers and the image of disabled people as part of the rest of us, was now also part of the company. . . . So I guess we felt . . . some of our people left, went to other companies, times had changed and it was just . . . it was time; it was time to move on."[69] Although Brightman felt that Apple had ingrained accessibility into its development process, Apple would forget its commitment to people with disabilities in a number of major ways in the following years.

Immediately after firing Worldwide Disability Solutions, Apple forced its Web site project Convomania to shut down. The site acted as a social gathering place for children with life-threatening diseases, providing connections to chat rooms and hosting mailing lists. Apple promised to allow the site to remain active under a slightly different name (Convonation) and to be run by volunteers, but it then threatened to sue when maintainers tried to make changes to the site, claiming that the company's property was being altered without consent. A spokesperson for Apple, Rhona Hamilton, admitted that the company had handled the situation poorly: "I think it was handled more from the corporate way of doing things than the way Apple is used to, but that seems to be happening a lot around here these days."[70] In spite of Apple's lack of support, the children and one adult volunteer were able to bring the site back to life on their own, rebuilding it from the ground up.

In addition to firing its own disability group, Apple also stopped supporting the Alliance for Technology Access. By 2000, Apple was no longer listed as a major supporter of the organization in its annual report.[71] IBM replaced Apple as the Alliance's main corporate supporter, starting with donations in the early 1990s and continuing to give throughout the decade and into the 2000s, with Microsoft joining their efforts by the late 1990s. Some of the ATA's original supporters also continued their contributions, such as IntelliTools and Pacific Bell.

By the late 1990s, disability and technology advocates such as Brand and Brightman had accomplished what they had set out to do almost two decades earlier. Basic accessibility features had become standard in personal computers, more people with disabilities had access to their own computers than ever before, and technological challenges such as the GUI had been at least partially conquered. However, the fight to make accessibility an intrinsic part of the computer technology development process still encountered situations where people with disabilities were overlooked in favor of other concerns. One of the most notable examples of the ongoing release of products that prevented people with disabilities from accessing them occurred with Microsoft's Internet Explorer 4 (IE4) in 1997. Although Internet Explorer 3 possessed accessibility features that made it compatible with the screen readers used by people with vision impairments, in a move to challenge the rising popularity of rival Internet browser Netscape Navigator by quickly pushing out a browser with new features, IE4 broke with Microsoft's own accessibility standards. The new version of the browser lacked Active Accessibility, Microsoft's software programming interface that allowed any application to communicate directly with adaptive devices, including screen readers, so that it could pass on useful information to the user. According to communication between members of the Microsoft Accessibility Team and the National Federation of the Blind, accessibility concerns in IE4 had been overridden in order to release the product quickly.[72] This is an example of the limitations inherent in posing market solutions to social problems; concerns other than accessibility trumped Microsoft's own standards and led to an inaccessible product.

Microsoft responded quickly to criticism from the blind community, however, and released a new version of IE4 a month later that fixed some of the accessibility problems, although screen-reading software still did not work correctly with it.[73] A year later, Bill Gates gave a speech at Microsoft's "Accessibility Day," admitting the error and promising to dedicate the company's attention to improving accessibility and preventing future problems.[74] This episode showed that even developers who seemed to have previously understood the needs of their diverse user base and instituted standards to allow for technological accommodation could ignore those needs when other corporate priorities loomed. The fight for personal computer accessibility—to allow all users to experience the benefits the technology can provide—still continues between developers and users with disabilities.

The Promises of Personal Computers

A t the close of the twentieth century, Raymond Kurzweil outlined his views on the future of computing technology in his book *The Age of Spiritual Machines.*[1] In this follow-up to his 1990 book, *The Age of Intelligent Machines,* Kurzweil used the history of computer development to predict a twenty-first century revolution in the way we understand humanity, intelligence, and our relationship to our bodies, a revolution with particular salience for people with disabilities. A 2004 documentary, *Freedom Machines,* offers a more grounded perspective on people with disabilities and technology, considering the different kinds of potential assistive technologies hold for people and the difficulties in acquiring such technologies.[2] These two perspectives on what technology means for people with disabilities who use it exemplify some of the tensions that run through this history of accessibility in the computer industry.

Kurzweil believed that computer technology has the potential to augment human intelligence and human bodies. He viewed this augmentation as a process that began with the computer's calculation capabilities: "Computers started out as extensions of our minds, and they will end up extending our bodies."[3] He thought that computer technology would evolve from performing calculations beyond human capacity to altering, improving, and ultimately replacing parts of the body. Eventually, according to Kurzweil, it would even replace the physical brain itself. He believed consciousness to be an emergent property residing in patterns of electrical and chemical activity and that one day people would figure out how to duplicate and transfer it to a digital form. Kurzweil anticipated that the ability of computer technology to extend human bodies would help humanity solve core biological problems, including that of mortality itself, and that these changes, only on the horizon in the late 1990s, would accelerate during the next century: "The twenty-first century will be different. The human species, along with the computational technology it created, will be able to solve age-old problems of need, if not desire, and will be in a position to change the nature of mortality in a postbiological future."[4] Computer technology and its derivatives, such as programmable nanotechnology, were the ultimate problem solvers in Kurzweil's utopian vision of the future.

The potential of computer technology, for Kurzweil, was a result of its superiority over biology. Created by humans, machines lacked the flaws and limitations of the body, especially those of the human brain. Following estimates of the exponential growth of the speed and processing power of computer technology, Kurzweil predicted that by 2020 a personal computer would be equivalent (in terms of speed and capacity) to a human brain. As this development occurred, technology would also increasingly be integrated into the body and brain. "We will enhance our brains gradually through direct connection with machine intelligence until such time that the essence of our thinking has fully migrated to the far more capable and reliable new machinery." Kurzweil saw this progression leading to a state of technology triumphing over, and perfecting upon, biology. By the end of the twenty-first century, he predicted that human intelligence would no longer be a function of biological neurons but of software connected to virtual bodies.[5]

Kurzweil's ideas for the future of computer technology and his foundation in developing accessible technologies came together in his methods for solving the problems people with disabilities face. His solution called for fixing the bodies of people with disabilities in order to allow them to function in a world not designed to meet their needs. Kurzweil provided a number of examples, both real and imagined, of the application of computer technology to accommodate disabled bodies. In a more contemporary example, he discussed the use of neural implants to alleviate Parkinson's patients' symptoms. These implants inhibited over-activation in the parts of the brain that cause the paralysis and stiffness that the disease brings on. Kurzweil also discussed the use of similar neural implants to treat people with tremor-causing diseases, such as cerebral palsy and multiple sclerosis. The linkage of implant and body allowed technology to directly affect the electrical impulses of the brain. "Increasingly, we are starting to combat cognitive and sensory afflictions by treating the brain and nervous system like the complex computational system that it is."[6] Machine and brain operate using the same physical principles, allowing technology to override instances where parts of the brain malfunction in specific ways.

Another type of neural implant to benefit people with disabilities that Kurzweil discussed is cochlear implants. Although the technology is highly contentious and considered by some Deaf activists as an attack on cultural Deafness, Kurzweil treated it unproblematically as a cure for a disability. He cited that, as of 1999, "About 10 percent of the formerly deaf persons who have received this neural replacement device are now able to hear and understand voices well enough that they can hold conversations using a normal telephone."[7] The cochlear implant acted the same for the deaf person as the neural implant that allowed the Parkinson's patient to move; it permitted the deaf person to function the same as a nondisabled person. Moreover, for Kurzweil, the cochlear implant essentially removed the disability itself, rendering its user "formerly deaf" and allowing them to operate the "normal" communication technology of the telephone.

These accessible technologies that, for Kurzweil, removed the handicaps of disability also had the potential to affect nondisabled users. Devices and features created with specialized uses for people with disabilities frequently diversified to include other kinds of uses. Many of these technologies improved usability and flexibility overall, allowing for more options in how users could interact with them, in the same way that computer technologies designed for people with disabilities found wider usefulness in increasing usability for everyone. Kurzweil believed this transference of specific design to general use would change technologies that erased disability into ones that would enhance humanity in general. Future neural implants, in particular, carried the potential to augment all human abilities: "Directly enhancing the information processing of our brain with synthetic circuits is focusing at first on correcting the glaring defects caused by neurological and sensory diseases and disabilities. Ultimately we will all find the benefits of extending our abilities through neural implants difficult to resist." This use of computer technology to alter human brains was not without its potential for abuse. In his predictions for the development of the technology, Kurzweil considered how it would change what it means to be human, for example, by granting people technological control over their emotions: "Once a technology is developed to overcome a disability, there is no way to restrict its use from enhancing normal abilities, nor would such restrictions necessarily be desirable. The ability to control our feelings will be just another one of those twenty-first-century slippery slopes."[8] Kurzweil's slippery slope worries about computer technology were similar to the case of the recreational use of pharmaceuticals—a technology intended to treat illness or disability being used by people to alter themselves in nonmedically prescribed ways.

Kurzweil's conception of the relationship between technology and disability is ultimately triumphalist and utopian. In many aspects, his discourse is similar to that of the disability activists and technology developers I examined. He hoped that computer technology would allow people with disabilities to have the opportunities that have been denied to them, the same opportunities available to people who are not disabled. By 2009, he was predicting that some of this equalizing would come to pass: "There is a growing perception that the primary disabilities of blindness, deafness, and physical impairment do not necessarily impart handicaps. Disabled persons routinely describe their disabilities as mere inconveniences. Intelligent technology has become the great leveler."[9] Like disability and technology advocates, Kurzweil hoped that computer technology would provide a more level playing field for people with disabilities by accommodating their needs. However, his focus on solving disability by fixing what is wrong with the body differed greatly from the stance of the activists, developers, and users I discussed. There is no attempt in Kurzweil's philosophy to change either social views or the built environment as a means of solving disability. For example, one of his predictions for future technologies—slated to begin use in 2009—was an orthotic walking machine for paraplegic people; these devices allowed people who would otherwise use a wheelchair to be able to climb stairs. The body must adapt to the environment here, not the other way around. Though he allowed for flexible uses of technology, he missed the possibilities of technology in redefining norms in such a way that disabled bodies are not seen as something that needs to be fixed in order for full participation in society to be possible. For him, the curb cut does not exist as a solution to problems of disability.

Kurzweil's view of fixing the bodies of people with disabilities contrasted with the solutions suggested by the social model of disability and by disability and technology activists. For Kurzweil, disability lay within individuals and was to be solved individually. What he called "handicaps" are the social dimension of disability, the limitations people with disabilities encounter in society. His solution was to correct the body, not society. Technology was the means to directly fix bodies and to indirectly allow for abilities the body lacks; that is, in his view, even before computer technology progressed to a point where it could physically repair the body, it would still be able to overcome disabilities by granting people those abilities their bodies are incapable of (as assistive technologies do today). The personal computer, however, erases many of the distinctions between a technology that changes the body to fit society and one that changes the social environment to accommodate bodies. It lies somewhere between a prosthetic limb and a building ramp; the personal computer is both a personal technology, granting its user new abilities, and a social one, acting as a portal to sites of social interaction. It is a mediator between the body and the world. Accessible personal computer technologies change what the individual is capable of and make possible social spaces where users with different kinds of abilities are all accommodated. The personal computer also acts to erase some of the distinctions between disabled and nondisabled by providing new abilities and augmentation for all users. For some people, the technology is more like Kurzweil's remediation; it overcomes the body's limitations. For others, it is more a mode of access to the public sphere. And for others still, the distinction between the two is meaningless; the personal computer provides for both.

Examples of some of the actual uses of accessible personal computer technology by people with disabilities and their meanings for users are shown in the 2004 documentary film, *Freedom Machines*.[10] The film, which aired on PBS, presents interviews with a number of people with disabilities talking about their experiences with assistive technology, along with disability activists—including Jackie Brand—pushing for greater availability and funding for such technologies. The technologies featured are mostly personal computer technologies and various types of wheelchairs. These people with disabilities, their families, and friends provide a perspective on the promise of personal computer technologies that is grounded in the everyday needs of people trying to participate in society in the ways they choose.

The film argues that technological accommodations are necessary for equity to be achieved. Standing in the way of access to these technologies is a lack of resources. As Brand explains, "It's a terribly frustrating thing, to look at something you know would change your life so enormously and be so powerful for you and to know it's not to be had, because you don't have the resources and this society has not decided that it's important enough for you to have."[11] Other people in the film echo the difficulties in acquiring the technologies that might benefit them or even in learning about the existence of such technologies in the first place. As one woman says, "I thought that with all this technology surely there was something out there that a visually impaired person could use. But I didn't realize it had been out for years and I just didn't know about it." A lack of information can keep technology out of the hands of the people it is intended to benefit. Social technologies are necessary to communicate to users knowledge on what technology exists and how to use it; they also provide a means for users to communicate their needs back to developers. With the activists I studied, the lines between these groups blurred, as many developers were also activists and many activists were users.

As it encourages greater access to technology, *Freedom Machines* also suggests that the best solutions to the problems of disability are based in universal design, where the social environment is made accessible to people with different needs instead of requiring people to adapt themselves to the environment via their own personal technologies. In the documentary, Rich Kjeldsen, an inventor of accessible computer interfaces, argues for the need for flexibility in technology to better accommodate use: "Right now we're in kind of an awkward stage, because technology is so rigid and fixed. And that's exactly what we're trying to do, is to make it more flexible, so that the technology adapts to us. And in that way, hopefully, we'll be able to open doors for everyone." For these users and developers most directly involved with accessible personal computer technology, the possibility of it truly acting as a leveler, of being one of the means of enacting equal rights, demonstrates its significance for people with disabilities. Although the filmmakers are aware of the difficulties in acquiring accessible technologies and in figuring out how to use them to meet individual needs, the film ultimately still proclaims a celebratory view of technology; the computer is presented as having the potential to allow new abilities and new forms of social interaction for people with disabilities. The promise of the personal computer, while perhaps impossible to fulfill, lies in the values embedded in it that allow it to be used and changed to fit the individual needs of different users and to be put to whatever uses they can imagine.

As I have traced the development of accessible personal computer technology, I have discovered a story about the struggle to fulfill the promise of the personal computer in order to benefit people with disabilities—people with whom the values embedded in the technology connect directly. Over time, the personal computer's values of openness, shared information, universality, and augmentation all came together in technologies that allowed people with disabilities access to new abilities as well as to the same technological features everyone seeks when using the computer. I observed the creation of technologies to accommodate the individual physical needs of people, but I also found work at the social level to fund these technologies and disseminate them, an approach that Kurzweil does not discuss. *Freedom Machines* provides a view of assistive technology and disability similar to the historical accounts I examined, a view grounded in the actual interactions of people with disabilities and their technology. My argument, however, extends beyond the film's perspective, which is still a fairly individual look at technology use. I consider the ways that accessible personal computer technologies are a necessary part of the enactment of civil rights and the possibility of full participation in society for people with disabilities. Only by examining the role technology plays in social equity can we fully understand the meaning the technology holds for the people who use it.

In order for personal computers to be used by people with different abilities, they had to be made accessible and flexible for various kinds of use. Developers pursued ideals similar to those of universal design in order to understand the different needs of users and to reimagine who counted as a "normal" user. Instead of averaging the needs of users and designing computer technology to fit an imagined, universal human being, universal design calls for a universality comprised of all possible differences. By understanding and accommodating the variety of user needs, technology can be universally usable. In the process of accessible computer technologies being developed, people with disabilities became the paradigmatic computer users. The technology directly accommodated their abilities and also frequently transformed features created for specialized purposes into features for general use by people who did not identify as disabled. By increasing usability to meet the needs of people with disabilities, personal computer technologies were made more flexible and usable overall, accommodating many other kinds of use as well.

The history I have told has been a mostly celebratory one. While technology developers did not build accessibility into the foundation of the personal computer development process as fully as activists hoped and while there remains a consistent lag between the release of a new technology and the updates that make it accessible, in many ways accessibility has still been generalized and made mainstream. Most computer companies have employees dedicated to promoting the development of accessibility for their products. Normalcy has come to be redefined, to an extent. When computer development takes into account the ideals of universal design and attempts to capture as many users as possible within the market share, who counts as imagined users is a far larger group than it was previously. Accessibility is in many ways now standard, and universal design is a buzzword in industry. Kurzweil was correct in that if accessibility and accommodating the needs of people with disabilities is addressed first, the usability of the technology to benefit all users will follow. By the end of the twentieth century, even when developers released technologies that did not take accessibility into account (when other corporate concerns trumped accessibility), an infrastructure of legislation, activists, and users was in place to patch the lack of accessibility. This network of laws, companies, organizations, and individuals which had developed since the

1970s worked together to implement accessibility in personal computer technology and fulfill some of its more utopian promises. The responsiveness to accessibility and the needs of people with disabilities changed during this history.

Unlike Kurzweil's transhumanist view of implanting the computer into the body in order to correct it and solve disability, what was historically required to bring people with disabilities into fuller participation in society was a political infrastructure that backed accessibility as an issue of equity and a change in corporate philosophy that saw people with disabilities as intended users and consumers. These changes in attitudes toward disability were a part of the cultural context within which the personal computer developed. The personal computer represents the possibility for changes in the meanings of disability and normalcy—a blurring of lines between categories that changes what is possible when assumptions about who counts as a user and how they might use technology are made to accommodate a multiplicity of needs.

Notes

INTRODUCTION: The Development of Accessible Personal Computer Technologies

1. Raymond C. Kurzweil, *The Age of Spiritual Machines: When Computers Exceed Human Intelligence* (New York: Viking, 1999), 174.
2. Ibid., 175.
3. Ibid., 177.
4. Fatima Vieira, "The Concept of Utopia," in *The Cambridge Guide to Utopian Literature,* ed. Gregory Claeys (Cambridge: Cambridge University Press, 2010), 9–10.

CHAPTER ONE: Disability Rights and Technology before the Personal Computer

1. I primarily draw on the work of four scholars in my analysis of disability rights in the United States: Joseph P. Shapiro, *No Pity: People with Disabilities Forging a New Civil Rights Movement* (New York: Three Rivers Press, 1994); Richard K. Scotch, *From Good Will to Civil Rights: Transforming Federal Disability Policy* (Philadelphia: Temple University Press, 1984); Kent Hull, *The Rights of Physically Disabled People* (New York: Avon Books, 1979); and Claire H. Liachowitz, *Disability as a Social Construct: Legislative Roots* (Philadelphia: University of Pennsylvania Press, 1988).
2. Scotch, *From Good Will to Civil Rights,* 9.
3. Shapiro, *No Pity,* 59–60.
4. Liachowitz, *Disability as a Social Construct,* 40.
5. Scotch, *From Good Will to Civil Rights,* 20.
6. Shapiro, *No Pity,* 64.
7. Ibid., 17–18.
8. Scotch, *From Good Will to Civil Rights,* 10.
9. Architectural Barriers Act of 1968, 42 U.S.C. § 4151 (1968).
10. Scotch, *From Good Will to Civil Rights,* 30.
11. Hull, *The Rights of Physically Disabled People,* 65–66.
12. Ibid., 67.
13. Quoted in ibid., 79.
14. Rehabilitation Act of 1973, Pub. L. No. 93-112, 87 Stat. 394, 29 (1973).
15. Scotch, *From Good Will to Civil Rights,* 52.
16. Ibid., 51 and 139.
17. Ibid., 60.

18. Ibid., 79.

19. Ibid., 84.

20. Ibid., 93.

21. Shapiro, *No Pity*, 67.

22. *Nondiscrimination on the Basis of Handicap in Programs and Activities Receiving or Benefiting from Federal Financial Assistance*, 42 Fed. Reg. 22676 (May 4, 1977).

23. Ibid., 22677–22678.

24. Ibid., 22676.

25. Rehabilitation Act of 1973, Pub. L. No. 93-112, 87 Stat. 394, 29 (1973).

26. Rehabilitation Act Amendments of 1974, Pub. L. No. 93-516, 88 Stat. 1617 (1974).

27. Education for All Handicapped Children Act of 1975, Pub. L. No. 94-142 (1975).

28. Shapiro, *No Pity*, 70.

29. Ibid., 73.

30. Hull, *The Rights of Physically Disabled People*, 22.

31. Ibid., 162.

32. Ibid., 164.

33. Scotch, *From Good Will to Civil Rights*, 164.

34. Hull, *The Rights of Physically Disabled People*, 14.

35. Shapiro, *No Pity*, 70–71.

36. Theodor D. Sterling et al., "Professional Computer Work for the Blind," *Communications of the ACM* 7, no. 4 (1964): 228–51.

37. Ibid., 228.

38. Ibid.

39. In this method of printing braille, the printer repeatedly printed a period in the same spot, causing the back of the paper to be embossed. This process would not create as permanent an imprint as a mechanical embosser could, but it would stay readable for some time if stored carefully.

40. Ibid., 229.

41. Ibid., 230.

42. Committee on Professional Activities of the Blind, *The Selection, Training, and Placement of Blind Computer Programmers* (n.p.: Association for Computing Machinery, 1966).

43. Ibid., 36.

44. Ibid., 7.

45. Ibid., 15. This aside is a part of a section on professional behavior for the potential employee.

46. Gordon Cummings, "Blind Programmer Questionnaire," *SIGCAPH Newsletter* no. 8 (1973): 4–13.

47. "SIG/SIC Functions," *SICCAPH Newsletter*, no. 5 (1971): 2.

48. "Bylaws for SIGCAPH," *SIGCAPH Newsletter*, no. 9 (1973): 7.

49. Robert A. J. Gildea, "Chairman's Message," *SIGCAPH Newsletter*, no. 10 (1974): 1.

50. Herbert S. Bright, "Letter to SIGCAPH," *SIGCAPH Newsletter*, no. 12 (1974): 7.

51. "SICCAPH Mid-Winter Meeting," *SIGCAPH Newsletter*, no. 6 (1972): 3.

52. "An Idea: Should We Drop the Braille Version of SIGCAPH Newsletter and Shift to Cassette Tapes Instead?," *SIGCAPH Newsletter*, no. 23 (1978): 3.

53. Kari Larsen, "Reg.: Drop Braille Version? NO!!! (SIGCAPH no. 23)," *SIGCAPH Newsletter*, no. 24 (1978): 13.

54. "The Shift to Tape Cassette," *SIGCAPH Newsletter*, no. 26 (1980): 2.

55. Murray Turoff, "Computerized Conferencing for the Deaf and Handicapped," *SIGCAPH Newsletter*, no. 16 (1975): 5.

56. Ibid.

57. "Conferencing System for Handicapped," *SIGCAPH Newsletter*, no. 19 (1976): 1.

58. Starr Roxanne Hiltz and Murray Turoff, *The Network Nation: Human Communication via Computer* (Reading, MA: Addison-Wesley, 1978). A second edition of the book was published in 1993.

59. Ibid., xxv.

60. Ibid., xxix.

61. Ibid., 169.

62. New Jersey Institute of Technology, "Research Activity . . . 'Computer Conferencing' . . . ," *SIGCAPH Newsletter*, no. 25 (1979): 12.

63. Ibid., 14.

64. Hiltz and Turoff, *The Network Nation*, 338.

65. Ibid., 173.

66. Emerson W. Pugh, *Building IBM: Shaping an Industry and Its Technology* (Cambridge, MA: MIT Press, 1995), chapter 2, "Origins of IBM."

67. Paul Ceruzzi, *A History of Modern Computing*, 2nd ed. (Cambridge, MA: MIT Press, 2003), 14 and 110. Seventy percent is the commonly used figure to describe IBM's control of the market; however, Emerson Pugh argues that this number was based on a limited conception of what constituted the computer industry. During their Justice Department lawsuit, IBM supplied figures that put their control of the market at 60 percent during the 1950s and under 40 percent by the 1970s (Pugh, *Building IBM*, 319).

68. IBM, "Think: A History of Progress: 1890s to 2001," 2008, www-03.ibm.com/ibm /history/interactive/ibm_history.pdf., accessed Aug. 11, 2012.

69. Annemarie Cooke, "A History of Accessibility at IBM," *Access World* 5, no. 2 (Mar. 2004). Online at www.afb.org/afbpress/pub.asp?DocID=aw050207, accessed Aug. 29, 2012; "Seventy Years of Enabling the Disabled," *Think*, no. 3 (1988): 43.

70. Claire Stegmann, "Handicapped? Not on the Job," *Think* (July–Aug. 1977): 42.

71. IBM, "IBM's focus on accessibility," 2008, www-03.ibm.com/able/product_acces sibility/ibmcommitment.html, accessed Aug. 11, 2012.

72. Stegmann, "Handicapped? Not on the Job," 42–47.

73. "Seventy Years of Enabling the Disabled," 43, and Pugh, *Building IBM*, 324.

74. Kathy Kafer, "A Fair Chance," *Think*, no. 3 (1988): 41.

75. "Seventy Years of Enabling the Disabled," 43. Though this article is from the late 1980s, the types of accommodations it describes are not specific to the personal computer but are the types of accommodations that any employer would need to enact to create a barrier-free workplace.

76. Quoted in A. N. Borno, "A Lifeline to Society," *Think* (Mar. 1972): 17.

77. Ibid.

78. IBM National Support Center for Persons with Disabilities, *Technology for Persons with Disabilities: An Introduction* (n.p.: IBM, 1990), 17.

79. "Seventy Years of Enabling the Disabled," 43.

80. Jay W. Spechler, *Reasonable Accommodation: Profitable Compliance with the Americans with Disabilities Act* (Delray Beach, FL: St. Lucie Press, 1996), 129.

CHAPTER TWO: **Early Personal Computer Accessibility, 1980–1987**

1. Peter A. McWilliams, *Personal Computers and the Disabled* (Garden City, NY: Garden Press, 1984), 14.

2. Dolores Hagen, *Microcomputer Resource Book for Special Education* (Reston, VA: Reston Publishing Co., 1984), 9.

3. Fred Turner, *From Counterculture to Cyberculture: Stewart Brand, the Whole Earth Network, and the Rise of Digital Utopianism* (Chicago: University of Chicago Press, 2006).

4. The Trace Research and Development Center on Communication, Control, and Computer Access for Handicapped Individuals at the University of Wisconsin–Madison is a major location for work on disabilities and computers in academia in the United States.

5. Gregg C. Vanderheiden, *Curbcuts and Computers: Providing Access to Computers and Information Systems for Disabled Individuals* (Madison, WI: Trace Research and Development Center, 1983), 5, www.eric.ed.gov/ERICWebPortal/detail?accno=ED289314.

6. Frank G. Bowe, *Personal Computers and Special Needs* (Berkeley, CA: Sybex, 1984), 133.

7. Ibid., 133–34. Bowe uses an argument here made by Richard Heddinger, a statistician and federal employee involved with filing the lawsuit against the D.C. Metro system in 1972.

8. Vanderheiden, *Curbcuts and Computers*, 5.

9. "Personal Computers Help the Handicapped: Johns Hopkins Rewards Inventors," *Creative Computing* 8, no. 3 (Mar. 1982): 54–55.

10. "Personal Computing for the Handicapped (National Contest)," *SIGCAPH Newsletter*, no. 29 (1981): 16.

11. Ibid., 17.

12. Ibid.

13. Paul L. Hazan, "Computing and the Handicapped: Guest Editor's Introduction," *Computer* 14, no. 1 (Jan. 1981): 9.

14. Harry Levitt, "A Pocket Telecommunicator for the Deaf," in *Proceedings of the Johns Hopkins First National Search for Applications of Personal Computing to Aid the Handicapped* (Los Angeles: IEEE Computer Society, 1981), 39.

15. "Personal Computers Help the Handicapped: Johns Hopkins Rewards Inventors," 54.

16. Mark B. Friedman, Gary Kiliany, Mark Dzmura, and Drew Anderson, "The EyeTracker Communication System," in *Proceedings of the Johns Hopkins First National Search*, 183.

17. Ibid., 185.

18. Robin L. Hight, "Lip-Reader Trainer: A Computer Program for the Hearing Impaired," in *Proceedings of the Johns Hopkins First National Search*, 4–5.

19. Joseph T. Cohn, "Microcomputer Augmentative Communication Devices," in *Proceedings of the Johns Hopkins First National Search*, 43–44.

20. Randy W. Dipner, "The Micro-Braille System," in *Proceedings of the Johns Hopkins First National Search*, 244–45.

21. Robert Stepp, "A Braille Word Processing System," in *Proceedings of the Johns Hopkins First National Search*, 202.

22. Sandra Jackson, Judy Maples Simmons, and Tony Wedig, "We Help More," in *Proceedings of the Johns Hopkins First National Search*, 59–60.

23. David L. Jaffe, "An Ultrasonic Head Position Interface for Wheelchair Control," in *Proceedings of the Johns Hopkins First National Search*, 142–43.

24. Paul Schwejda and Judy McDonald, "Adapting the Apple for Physically Handicapped Users: Two Different Solutions," in *Proceedings of the Johns Hopkins First National Search*, 53–54, and "Personal Computers Help the Handicapped: Johns Hopkins Rewards Inventors," 55.

25. Raymond C. Kurzweil, "Kurzweil Reading Machine for the Blind," in *Proceedings of the Johns Hopkins First National Search*, 236.

26. Quoted in McWilliams, *Personal Computers and the Disabled*, 56.

27. Ibid., 59.

28. Bowe, *Personal Computers and Special Needs*, 123, 124.

29. Ibid., 112.

30. Ibid., 114.

31. McWilliams, *Personal Computers and the Disabled*, 293.

32. Ibid., 88.

33. Ibid., 90.

34. See ibid., 297 and 300, for descriptions of Large Type and PC-LENS.

35. Ibid., 63.

36. Ibid., 309 and 301. Portable speech synthesizers listed by McWilliams ran from $150 (the Vocaid) to almost $3,000 (the Phonic Ear Vois).

37. Ibid., 81.

38. Bowe, *Personal Computers and Special Needs*, 86.

39. McWilliams, *Personal Computers and the Disabled*, 83–85.

40. Bowe, *Personal Computers and Special Needs*, 129.

41. Ibid., 18.

42. Ibid., 62 and 64.

43. Ibid., 21.

44. Michael J Silva et al., Membrane Computer Keyboard and Method, US Patent 5,450,078, filed Oct. 8, 1992, and issued Sept. 12, 1995.

45. Ibid., 1.

46. Ibid., 2.

47. James H. Heller, David Salisbury, and Judith C. Lapadat, "The Unicorn Model 1 Keyboard as a Rehabilitation Tool," in *Computer Technology for the Handicapped: Proceedings from the 1984 Closing The Gap Conference*, ed. Michael Gergen and Dolores Hagen (Henderson, MN: Closing the Gap, 1984), 68–70.

48. Ibid., 70.

49. Ibid., 69.

50. Shoshana Brand later changed her name to Judith, when she was an adult. For the sake of historical accuracy, I refer to her as Shoshana during the time period that that was her name.

51. Most of this history of the Brands' personal lives comes from Jacquelyn Brand, "Parent Advocate for Independent Living, Founder of the Disabled Children's Computer Group and the Alliance for Technology Access," an oral history conducted in 1998–99 by Denise Sherer Jacobson, in *Builders and Sustainers of the Independent Living Movement in Berkeley,* vol. 5, Regional Oral History Office, The Bancroft Library, University of California, Berkeley, 2000.

52. Ibid., 23, 27–28.

53. Ibid., 25–26.

54. Ibid., 37, 45.

55. Ibid., 52.

56. Mary Lester, "Grant Writer for the Early Center for Independent Living in Berkeley, 1974–1981," an oral history conducted in March 2000 by Susan O'Hara Jacobson, in *Builders and Sustainers of the Independent Living Movement in Berkeley,* vol. 1, Regional Oral History Office, Bancroft Library, University of California, Berkeley, 2000, 114–15.

57. Jacquelyn Brand, "Families Working Together," *Exceptional Parent* 15, no. 6 (Oct. 1985): 17–18.

58. Jacquelyn Brand, "Parent Advocate for Independent Living," 54.

59. Ibid., 55.

60. Hagen, *Microcomputer Resource Book for Special Education,* 2.

61. Apple Computer, *Simplicity Is the Ultimate Sophistication: Introducing Apple II, the Personal Computer,* Computer History Museum, www.computerhistory.org/brochures/full_record.php?iid=doc-43729572aadaf.

62. Hagen, *Microcomputer Resource Book for Special Education,* 4.

63. Quoted in Harvey Pressman, "The National Special Education Alliance: Applying Microcomputer Technology to Benefit Disabled Children and Adults" (Cupertino, CA: National Special Education Alliance, Apple Computer, Inc., 1987), 10, box 1, folder 1, Coll. BANC MSS 99/248c, Bancroft Library, University of California, Berkeley.

64. Jane Ferrell, "Computers Help the Disabled Get an Equal Chance," *San Francisco Examiner,* Mar. 24, 1985, D3.

65. Jacquelyn Brand, "Families Working Together," *Exceptional Parent,* Oct. 1985, 18.

66. Jackie Brand, "Assistive Technology Oral History Project," interview with Chauncy Rucker, Nov. 1, 2007, http://atoralhistory.uconn.edu/podcasts/brand.php.

67. Ibid., 56, and Pressman, "The National Special Education Alliance," 5.

68. Jacquelyn Brand, "The Disabled Children's Computer Group," *Exceptional Parent,* Oct. 1985, 16.

69. Jacquelyn Brand, "Parent Advocate for Independent Living," 56 and 58.

70. "Developing a Parent/Community Technology Resource Center," *Closing the Gap,* Apr. 12, 1986, 1.

71. Steve Brand, "The President's Message," *DCCG 1985 Annual Report,* n.p.

72. Jacquelyn Brand, "Parent Advocate for Independent Living," 56–57.

73. "Developing a Parent/Community Technology Resource Center," 1.

74. Pressman, "The National Special Education Alliance," 5–6.

75. Jacquelyn Brand, "Families Working Together," 17.

76. Ibid., 18.

77. Ibid., 17.

78. Jack Kenny, "Bridging the sensory divide," *TES Magazine,* Oct. 16, 1998, www.tes .co.uk/article.aspx?storycode=79600.

79. DCCG event calendars, May–June 1989 and Nov–Dec 1991, box 1, folder 7, Coll. BANC MSS 99/185c, Bancroft Library, University of California, Berkeley.

80. Jacquelyn Brand, "Parent Advocate for Independent Living," 58.

81. Street Electronics was bought by another company. Telesensory was bought then declared bankruptcy. Kurzweil sold his company to Xerox; it became Nuance, which makes Dragon NaturallySpeaking. Words+ was bought by Simulations Plus.

82. McWilliams, *Personal Computers and the Disabled,* 49.

83. Gregg C. Vanderheiden, *White Paper: Access to Standard Computers, Software, and Information Systems by Persons with Disabilities,* version 2.0 (Madison, WI: Trace Research and Development Center, 1985), 17, www.eric.ed.gov/ERICWebPortal/detail?accno=ED280257.

84. A successful example of IBM's dominance over innovation was in the lack of change in punch-card technology between the 1930s and 1960s. See Paul Ceruzzi, *A History of Modern Computing,* 2nd ed. (Cambridge, MA: MIT Press, 2003), 111.

85. Ceruzzi, *A History of Modern Computing,* 171.

86. Ibid., 272.

87. Ibid., 252, 277–79.

88. Ibid., 304, and Jeffrey R. Yost, *The Computer Industry* (Westport, CT: Greenwood, 2005), 183.

89. IBM National Support Center for Persons with Disabilities, *Technology for Persons with Disabilities: An Introduction* (IBM, 1990), 16. This booklet was one of the files available for users to download from the bulletin board.

90. Ibid., preface.

91. "National Support Center; a Service of IBM," *Exceptional Parent,* special issue, *8th Annual Computer Technology Directory,* Nov. 1, 1990, 2.

92. Jay W. Spechler, *Reasonable Accommodation: Profitable Compliance with the Americans with Disabilities Act* (Delray Beach, FL: St. Lucie Press, 1996), 129.

93. Kathy Kafer, "A Fair Chance," *Think,* no. 3 (1988): 44.

94. "IBM Independence Series" brochure, included with IBM National Support Center for Persons with Disabilities, *Technology for Persons with Disabilities: An Introduction.*

95. McWilliams, *Personal Computers and the Disabled,* 129.

96. Steven Levy, *Insanely Great: The Life and Times of Macintosh, the Computer That Changed Everything* (New York: Penguin, 1994), 134.

97. Ibid., 137.

98. Ibid., 109.

99. Ibid., 121.

100. Bowe, *Personal Computers and Special Needs,* 85.

101. Levy, *Insanely Great*, 194.

102. Ibid., 223.

103. Most of the details of Brightman's history here come from Alan Brightman, "Assistive Technology Oral History Project," interview with Chauncy Rucker, Mar. 13, 2008, http://atoralhistory.uconn.edu/podcasts/Brightman.php.

104. Ibid.

105. Ibid.

106. Diane Divoky, "Apple Sponsors a New Alliance for Disabled Computer Users," Classroom Computer Learning, Oct. 1987, 46–49.

107. Brightman, "Assistive Technology Oral History Project."

108. Apple Special Education, "Access," VHS, 1986.

109. Alan Brightman, "Microcomputers and Special Education: Lessons from Unreasonable People," in *Computer Technology for the Handicapped: Proceedings from the 1985 Closing the Gap Conference*, ed. Michael Gergen and Dolores Hagen (Henderson, MN: Closing the Gap, 1985), 1–6.

110. Jacquelyn Brand, "Parent Advocate for Independent Living," 60.

111. Meeting notes, DCCG Steering Committee and Board Meeting, Jan. 23, 1986, box 1, folder 3, Coll. BANC MSS 99/185c, Bancroft Library, University of California, Berkeley. I believe this is the same meeting where Brightman demonstrated the accessibility problems with the Macintosh to its engineers.

112. Alan Brightman to Jackie Brand, Mar. 7, 1986, box 1, folder 4, Coll. BANC MSS 99/185c, Bancroft Library, University of California, Berkeley.

113. Quoted in Alvaro Delgado, "Computers + disabled = hope," *West County Times*, May 22, 1986, 1B.

CHAPTER THREE: **Corporate Philanthropy and the National Special Education Alliance**

1. I use the names NSEA or ATA depending on the historical time I am discussing. I also use the term Alliance to refer to the organization in general, as this nickname was used throughout the group's history.

2. Jacquelyn Brand, "Parent Advocate for Independent Living, Founder of the Disabled Children's Computer Group and the Alliance for Technology Access," an oral history conducted in 1998–99 by Denise Sherer Jacobson in *Builders and Sustainers of the Independent Living Movement in Berkeley*, vol. 5, Regional Oral History Office, Bancroft Library, University of California, Berkeley, 2000, 57–58.

3. Robert Dodds Glass, "Partners in the Promise of Technology: An Historical Analysis and Impact Description of the Alliance for Technology Access," PhD diss., School of Education, University of Louisville, Kentucky, 1992, 54. A copy of this dissertation can be found in box 1, folder 2, Coll. BANC MSS 99/248c, Bancroft Library, University of California, Berkeley. Glass wrote his dissertation while working for the Alliance. He served on the Alliance Planning Team of the NSEA. When the organization changed its name to the Alliance for Technology Access, he became a board member and the assistant to the executive director. His research included materials from the ATA and interviews he conducted with various members.

4. Quoted in Harvey Pressman, "National Special Education Alliance," *Exceptional Parent* 17, no. 7 (1987): 12–18, 21–22.

5. Harvey Pressman, "The National Special Education Alliance: Applying Microcomputer Technology to Benefit Disabled Children and Adults" (Cupertino, CA: National Special Education Alliance, Apple Computer, Inc.), 1987, box 1, folder 1, Coll. BANC MSS 99/248c, Bancroft Library, University of California, Berkeley, 2.

6. Ibid., 22.

7. Ibid., 21.

8. Ibid.

9. Jacquelyn Brand, "From the National Special Education Alliance," *The DCCG Report: Rose Street Edition July, 1987–June, 1989*, 8, box 1, folder 2, Coll. BANC MSS 99/185c, Bancroft Library, University of California, Berkeley, 8.

10. Quoted in Diane Divoky, "Apple Sponsors a New Alliance for Disabled Computer Users," *Classroom Computer Learning*, Oct. 1987, 46.

11. Quoted in Owen W. Linzmayer, *Apple Confidential 2.0: The Definitive History of the World's Most Colorful Company* (San Francisco: No Starch Press, 2004), 82.

12. Stephen Wozniak and Gina Smith, *iWoz: Computer Geek to Cult Icon: How I Invented the Personal Computer, Co-founded Apple, and Had Fun Doing It* (New York: W. W. Norton and Co., 2006), 133.

13. Steven Levy, *Insanely Great: The Life and Times of Macintosh, the Computer That Changed Everything* (New York: Penguin, 1994), 122.

14. Ibid., 180, 224–25.

15. Ibid., 121. For Raskin, the clashes with Steve Jobs were both ideological and personal.

16. *Technology-Related Assistance for Individuals with Disabilities Act of 1988: Hearings on H.R. 4904, before the Subcommittee on Select Education of the Comm. on Education and Labor*, 100th Cong (1988) (statement of James Johnson, Director of Government Affairs, Apple Computer, Inc.), 54.

17. "What Do You Do When the World Tells You, 'That's All You Can Do?,'" Apple NSEA brochure, 1988, box 2, folder 6, Coll. BANC MSS 99/248c, Bancroft Library, University of California, Berkeley.

18. Glass, "Partners in the Promise of Technology," 54.

19. Jackie Brand, "Assistive Technology Oral History Project," interview with Chauncy Rucker, Nov. 1, 2007, http://atoralhistory.uconn.edu/podcasts/brand.php.

20. Pressman, "The National Special Education Alliance: Applying Microcomputer Technology," 1.

21. Ibid., 3.

22. Ibid., 2, 36.

23. Quoted in Divoky, "Apple Sponsors a New Alliance," 46–49.

24. Alan Brightman, "Assistive Technology Oral History Project," interview with Chauncy Rucker, Mar. 13, 2008, http://atoralhistory.uconn.edu/podcasts/Brightman.php.

25. Pressman, "The National Special Education Alliance: Applying Microcomputer Technology," 32–33.

26. Linzmayer, *Apple Confidential 2.0*, 147.

27. Ibid., 148.

28. Jacquelyn Brand, "Parent Advocate for Independent Living," 64.

29. Jane Ferrell, "Computers Help the Disabled Get an Equal Chance," *San Francisco Examiner*, Mar. 24, 1985, D3.

30. Pressman, "The National Special Education Alliance: Applying Microcomputer Technology," 32–33.

31. Jacquelyn Brand, "Parent Advocate for Independent Living," 64.

32. Glass, "Partners in the Promise of Technology," 55.

33. Quoted in Pressman, "The National Special Education Alliance: Applying Microcomputer Technology," 15.

34. Ibid., 12.

35. Quoted in Pressman, "National Special Education Alliance," 17.

36. "What Do You Do When the World Tells You, 'That's All You Can Do?'"

37. Glass, "Partners in the Promise of Technology," 57.

38. Pressman, "National Special Education Alliance," 18, 21.

39. Pressman, "The National Special Education Alliance: Applying Microcomputer Technology," 35.

40. Ibid.

41. Glass, "Partners in the Promise of Technology," 59.

42. Ibid., 61.

43. Ibid., 65.

44. Jacquelyn Brand, Form 1023–Application for Recognition of Exemption under Section 501(c)(3) of the Internal Revenue Code, 4/4/89, box 1, folder 3, Coll. BANC MSS 99/248c, Bancroft Library, University of California, Berkeley.

45. Glass, "Partners in the Promise of Technology," 67.

46. Ibid., 71–72.

47. Jacquelyn Brand, "Parent Advocate for Independent Living," 67.

48. Glass, "Partners in the Promise of Technology," 68–69.

49. Ibid., 72–73.

50. Jacquelyn Brand, Form 1023.

51. Jacquelyn Brand, "From the National Special Education Alliance," 8, and Glass, "Partners in the Promise of Technology," 136.

52. Kate Sefton, "From the Director," *The DCCG Report: Rose Street Edition July, 1987–June, 1989*, box 1, folder 2, Coll. BANC MSS 99/185c, Bancroft Library, University of California, Berkeley, 3.

53. Disabled Children's Computer Group, "DCCG Programs and Services," *The DCCG Report: Rose Street Edition July, 1987–June, 1989*, box 1, folder 2, Coll. BANC MSS 99/185c, Bancroft Library, University of California, Berkeley, 4–5.

54. DCCG event calendars, May–June 1989 and Nov.–Dec. 1991, box 1, folder 7, Coll. BANC MSS 99/185c, Bancroft Library, University of California, Berkeley.

55. Linda De Lucchi, "From the President," *The DCCG Report: Rose Street Edition July, 1987–June, 1989*, box 1, folder 2, Coll. BANC MSS 99/185c, Bancroft Library, University of California, Berkeley, 2.

56. Harvey Pressman, "Jackie Brand: DCCG," *Exceptional Parent* 17, no. 7 (1987): 57.

57. Pressman, "The National Special Education Alliance: Applying Microcomputer Technology," 7.

58. Sara Hirsch, "Toy Lending Library," *Bay View* 16, no. 5 (Feb. 1988): 10, The Junior League of Oakland-East Bay, box 1, folder 7, Coll. BANC MSS 99/185c, Bancroft Library, University of California, Berkeley.

59. Pressman, "The National Special Education Alliance: Applying Microcomputer Technology," 8.

60. IBM National Support Center for Persons with Disabilities, *Technology for Persons with Disabilities: An Introduction* (n.p.: IBM, 1990), preface.

61. Ibid., 11.

62. Carl Friedlander, ed., "Reduced Cost Microcomputers Available," *SIGCAPH Newsletter,* no. 39 (1988): 5–6.

63. "National Support Center; a Service of IBM," *Exceptional Parent,* special issue, *8th Annual Computer Technology Directory,* Nov 1, 1990, 2.

64. Friedlander, "Reduced Cost Microcomputers Available," 5.

65. Ibid.

66. Ibid., 5–6.

67. Annemarie Cooke, "A History of Accessibility at IBM," *Access World* 5, no. 2 (Mar. 2004), www.afb.org/afbpress/pub.asp?DocID=aw050207.

68. Ibid., and Kathy Kafer, "A Fair Chance," *Think,* no. 3 (1988): 41.

69. Quoted in Cooke, "A History of Accessibility at IBM."

70. "User's Guide for AccessDOS," IBM, last modified Jan. 21, 1997, ftp://ftp.software .ibm.com/sns/accessd.zip.

71. Ibid., and "IBM Independence Series" brochure, included with IBM National Support Center for Persons with Disabilities, *Technology for Persons with Disabilities.*

72. Diane Divoky, "Curb Cuts for Computers," *Classroom Computer Learning,* Oct. 1987, 48.

73. Kafer, "A Fair Chance," 45.

74. IBM National Support Center for Persons with Disabilities, *Technology for Persons with Disabilities,* 15.

75. Kafer, "A Fair Chance," 45.

76. Ibid., 41.

77. Curtis Chong, "Correspondence on the GUI Problem," *Computer Science Update,* National Federation of the Blind, Summer 1994, http://cd.textfiles.com/nfbfiles/nfbcs/CS9406 .TXT.

78. "IBM Independence Series" brochure.

79. Jay W. Spechler, *Reasonable Accommodation: Profitable Compliance with the Americans with Disabilities Act* (Delray Beach, FL: St. Lucie Press, 1996), 131.

80. Joseph J. Lazzaro, *Adapting PCs for Disabilities* (Reading, MA: Addison-Wesley, 1996), 104.

81. Frank R. Adams, Hubert Crepy, David Jameson, and James Thatcher, "IBM Products for Persons with Disabilities," in *GLOBECOM '89: IEEE Global Telecommunications Conference & Exhibition, Dallas, Texas, November 27–30, 1989, "Communications Technology for*

the 1990s and Beyond"; Conference Record (New York: Institute of Electrical and Electronics Engineers, 1990), 982–84.

82. Peter A. McWilliams, *Personal Computers and the Disabled* (Garden City, NY: Garden Press, 1984), 14 and 56, and Frank G. Bowe, *Personal Computers and Special Needs* (Berkeley, CA: Sybex, 1984), 19.

83. IBM National Support Center for Persons with Disabilities, *Technology for Persons with Disabilities,* 25.

84. "IBM VoiceType Simply Speaking Brings Speech Recognition Technology to Home, School, and Mobile Office," IBM Software Announcement, Letter Number 296-434, Oct. 29, 1996, available www-304.ibm.com/jct01003c/cgi-bin/common/ssi/ssialias?infotype =an&subtype=ca&htmlfid=897/ENUS296-434&appname=xldata&language=enus.

85. Lazzaro, *Adapting PCs for Disabilities,* 80.

86. "IBM Independence Series" brochure.

CHAPTER FOUR: **The Growth of Disability Rights and Accessible Computer Technologies**

1. Joseph P. Shapiro, *No Pity: People with Disabilities Forging a New Civil Rights Movement* (New York: Three Rivers, 1994), 85.

2. The word *Deaf* is capitalized in this manner to refer to the identity group.

3. Ibid., 74.

4. "Flyer distributed at the March 1, 1988 Rally," Gallaudet University, accessed Aug. 29, 2012, www.gallaudet.edu/Gallaudet_University/About_Gallaudet/DPN_Home/Issues /Related_Documents/RallyFlyers.html.

5. All letters can be found at www.gallaudet.edu/Gallaudet_University/About_Gallau det/DPN_Home/Issues/Letters_of_Support/.

6. Paul Simon, Paul Simon to Greg Hlibok, Washington, DC, Mar. 10, 1988, in Letters of Support, Gallaudet University, www.gallaudet.edu/Gallaudet_University/About_Gallaudet /DPN_Home/Issues/Letters_of_Support/Senator_Paul_Simon.html.

7. Shapiro, *No Pity,* 77.

8. President's Council on Deafness, "Position of the Students, Faculty and Staff of Gallaudet University," Gallaudet University, 79, accessed Aug. 29, 2012, www.gallaudet.edu /Gallaudet_University/About_Gallaudet/DPN_Home/Issues/Related_Documents/PCD _Demands.html.

9. Shapiro, No Pity, 78.

10. Ibid., 80.

11. Quoted in Molly Sinclair and Eric Pianin, "Protest May Imperil Gallaudet Funding: Some Members of Congress Back Movement for Deaf President," *Washington Post,* Mar. 9, 1988, A1.

12. Shapiro, *No Pity,* 83.

13. Ibid., 85.

14. Americans with Disabilities Act of 1988: Joint Hearings on S. 100-926, before the Subcommittee on the Handicapped of the Comm. on Labor and Human Resources United States Senate and the Subcommittee on Select Education of the Comm. on Education and Labor House of Representatives, 100th Cong. (1988), 36.

15. Tom Harkin, "A View from Capitol Hill," *PC/Computing,* July 1989, 91.

16. Technology-Related Assistance for Individuals with Disabilities Act of 1988, Pub. L. No. 100-407 (1988), Section 3.

17. Ibid., Section 2a.

18. Harkin, "A View from Capitol Hill," 91.

19. Technology-Related Assistance for Individuals with Disabilities Act of 1988, Section 2a.

20. Ibid., Section 2b.

21. *Technology-Related Assistance for Individuals with Disabilities Act of 1988: Hearings on H.R. 4904, before the Subcommittee on Select Education of the Comm. on Education and Labor,* 100th Cong (1988) (statement of James Johnson, Director of Government Affairs, Apple Computer, Inc.).

22. Ibid., 57.

23. Ibid., 56, 57.

24. "1992 Program Impact Report: Redefining Human Potential: The Partners, Progress and Promise of the Alliance for Technology Access," Foundation for Technology Access, box 1, folder 5, Coll. BANC MSS 99/248c, Bancroft Library, University of California, Berkeley.

25. *ATA Perspectives* 2 [no date], newsletter of the Foundation for Technology Access, Albany, California, edited by Jackie Brand, Mary Lester, and Mary Lou Sumberg, box 2, folder 6, Coll. BANC MSS 99/248c, Bancroft Library, University of California, Berkeley, 4.

26. National Council on the Handicapped, *Toward Independence: An Assessment of Federal Laws and Programs Affecting Persons with Disabilities—with Legislative Recommendations* (Washington, DC: National Council on the Handicapped, 1986).

27. Rehabilitation Act Amendments of 1984, Pub. L. No. 98-221, 98 Stat. 26 (1984), 142.

28. National Council on the Handicapped, *Toward Independence,* 12.

29. Ibid., 18.

30. Ibid., 19.

31. Ibid., 20.

32. National Council on the Handicapped, *On the Threshold of Independence: Progress on Legislative Recommendations from "Toward Independence"* (Washington, DC: National Council on the Handicapped, 1988).

33. Ibid., xiii, xviii, 4, 23.

34. Shapiro, *No Pity,* 108.

35. Ibid., 118.

36. Americans with Disabilities Act of 1988: Joint Hearings on S. 100-926.

37. Ibid., 4.

38. Ibid.

39. Ibid., 12.

40. Ibid., 86.

41. Shapiro, *No Pity,* 117.

42. Ibid., 114.

43. National Council on Disability, *Equality of Opportunity: The Making of the Americans with Disabilities Act* (Washington, DC: National Council on Disability, 2010).

44. Ibid., 68–69.

45. Americans with Disabilities Act of 1988: Joint Hearings on S. 100-926, 91.

46. National Council on Disability, *Equality of Opportunity,* 59–60.

47. Shapiro, *No Pity,* 118.

48. Ibid., 119.

49. National Council on Disability, *Equality of Opportunity,* 79.

50. Ibid., 81.

51. Ibid., 82.

52. Ibid., 80, 83.

53. Ibid., 146.

54. *ATA Perspectives* 1 (Aug. 1990), newsletter of the Foundation for Technology Access, Albany, California, edited by Jackie Brand, Mary Lester, and Mary Lou Sumberg, box 2, folder 6, Coll. BANC MSS 99/248c, Bancroft Library, University of California, Berkeley, 7.

55. Robert Dodds Glass, "Partners in the Promise of Technology: An Historical Analysis and Impact Description of the Alliance for Technology Access," PhD diss., School of Education, University of Louisville, Kentucky, Apr. 1992, box 1, folder 2, Coll. BANC MSS 99/248c, Bancroft Library, University of California, Berkeley, 136.

56. *ATA Perspectives* 1, p. 4.

57. Harvey Pressman, "When Is Different Really the Same?," CompuCID newsletter, no. 1, spring 1990, box 1, folder 5, Coll. BANC MSS 99/185c, Bancroft Library, University of California, Berkeley, 3–4.

58. Lauren Terrazzano, "Sights Are High at CompuCID Sites," CompuCID newsletter, no. 2, fall 1990, box 1, folder 5, Coll. BANC MSS 99/185c, Bancroft Library, University of California, Berkeley, 2.

59. *ATA Perspectives* 2, p. 4.

60. Bob Glass, "1994 Program Impact Report: Redefining Human Potential: The Partners, Progress and Promise of the Alliance for Technology Access," box 1, folder 5, Coll. BANC MSS 99/248c, Bancroft Library, University of California, Berkeley.

61. Beth Smith et al., "Real People, Real Technology, Real Solutions," *Exceptional Parent* 24, no. 11 (Nov. 1994).

62. Glass, "Partners in the Promise of Technology," 81, 83.

63. Alan Brightman, "Assistive Technology Oral History Project," interview with Chauncy Rucker, Mar. 13, 2008, http://atoralhistory.uconn.edu/podcasts/Brightman.php.

64. Ibid., 93–94.

65. Ibid., 79.

66. Ibid., 78.

67. Curtis Chong, "Correspondence on the GUI Problem," Computer Science Update, National Federation of the Blind, summer 1994, accessed Aug. 29, 2012, http://cd.textfiles.com/nfbfiles/nfbcs/CS9406.TXT.

68. Ibid.

69. *ATA Perspectives* 2, p. 3. The FTA was the Foundation for Technology Access, the original name of the nonprofit organization that ran the Alliance. The FTA name was dropped in 1994, and the organization has been referred to as the ATA since.

70. Ibid., 3.

71. "CES to get Computer Learning Lab," Cherokee County Herald, Oct. 23, 1991, 6A, accessed Aug. 29, 2012, http://news.google.com/newspapers?id=T88vAAAAIBAJ&sjid=X j4DAAAAIBAJ&pg=6859%2C1561202.

72. Ibid.

73. Some articles on the Mattel Family Learning Program from the late 1990s state that the ATA did not join until 1994. This statement appears to have been an error that was perpetuated. In a copy of Russ Holland, Tom Morales, Regina Rodman, Dave Grass, and Kathy Perini, "The Mattel Family Learning Program—An Innovative Community Partnership," *Proceedings of the Technology and Persons with Disabilities Conference 1999*, members of the project from the ATA and Mattel wrote the following aside after the incorrect 1994 date: "didn't our relationship with them begin prior to this?" (www.csun.edu/~hfdss006/conf /1999/proceedings/session0100.htm).

74. *ATA Perspectives* 2, p. 6.

75. *ATA Perspectives* 2, p. 5.

76. Holland et al., "The Mattel Family Learning Program."

77. Kim Burruss, CSUN's Child LAB Receives Gift of New Computers, California State University Northridge, Press Release, Sept. 22, 1998, www.csun.edu/~hfoa0102/press _releases/fall98/lab.html.

CHAPTER FIVE: **Accessibility and Software Applications in the 1990s**

1. Bettye Rose Connell et al., "The Principles of Universal Design," North Carolina State University, Center for Universal Design, Raleigh, 1997, www.ncsu.edu/www/ncsu/design /sod5/cud/about_ud/udprinciplestext.htm.

2. Martin Campbell-Kelly, *From Airline Reservations to Sonic the Hedgehog: A History of the Software Industry* (Cambridge, MA: MIT Press, 2003), 15–16, 234.

3. Lawrence H. Boyd, Wesley L. Boyd, and Gregg C. Vanderheiden, "The Graphical User Interface Crisis: Danger and Opportunity," Trace Center, Sept. 1990, www.eric.ed.gov /ERICWebPortal/detail?accno=ED333687.

4. Ibid., 4.

5. Ibid., 5.

6. Quoted in Steven Levy, *Insanely Great: The Life and Times of Macintosh, the Computer That Changed Everything* (New York: Penguin, 1994), 58.

7. James Thatcher, "Problems and Challenges of the Graphical User Interface," *Braille Monitor* 37, no. 1 (Jan. 1994), http://nfb.org/legacy/bm/bm94/brlm9401.htm.

8. "Impact!: Working Documents," Section 5.3, Spring 1991, box 2, folder 3, Coll. BANC MSS 99/248c, Bancroft Library, University of California, Berkeley.

9. Ibid., Section 5.3.1.3. I have copied Matvy's original spelling. I have, however, added spaces in between words.

10. Ibid., Section 5.3.1.1.

11. Ibid., Section 5.3.1.3.

12. People with certain learning disabilities affecting their ability to understand metaphors also struggled with the GUI's use of a desktop metaphor. They, too, would need to use screen-reading technology as a way to translate the symbolic, graphical representations

of icons into labels that described literally what actions the computer would take when something was clicked on.

13. Thatcher, "Problems and Challenges of the Graphical User Interface."

14. Herb Brody, "The Great Equalizer: PCs Empower the Disabled," *PC/Computing*, 2, no. 7 (July 1989): 84.

15. Thatcher, "Problems and Challenges of the Graphical User Interface."

16. Boyd et al., "The Graphical User Interface Crisis," 8–12.

17. Linda Wahl and Paul Hendrix, "Access to Computer-Based Telecommunications for People with Disabilities," 35, Disabled Children's Computer Group, 1993, box 1, folder 9, Coll. BANC MSS 99/185c, Bancroft Library, University of California, Berkeley.

18. Ibid.

19. Greg Lowney, "Message from Microsoft," *Computer Science Update*, National Federation of the Blind, Summer 1994, accessed Aug. 29, 2012, http://cd.textfiles.com/nfbfiles/nfbcs/CS9406.TXT.

20. Kenneth Frasse, "GUI Access: A Comparison of Screen-Readers (Part I)," Access Review 2, no. 2 (1997), www.nyise.org/whatsnew/review.txt.

21. Campbell-Kelly, *From Airline Reservations,* 277 and 279.

22. Ibid., 292.

23. Glenn Rifkin, "Competing through Innovation: The Case of Broderbund," *Strategy and Business,* no. 11 (1998): 48–50. By 1998, *Myst* had sold four million copies.

24. Ibid., 50.

25. Russ Holland, Tom Morales, and Mary Lester, *Alliance for Technology Access 1996 Impact Survey and Report* (San Rafael, CA: Alliance for Technology Access), 7, box 1, folder 6, Coll. BANC MSS 99/248c, Bancroft Library, University of California, Berkeley.

26. The Alliance for Technology Access, *Computer Resources for People with Disabilities: A Guide to Exploring Today's Assistive Technology* (Alameda, CA: Hunter House, 1994), 41.

27. Jacquelyn Brand, Bridgett Perry (ATA), Eric Winkler, Kyle Hart (Brøderbund Software), "Alliance for Technology Access and Brøderbund Software Join to Raise Awareness of Software Accessibility Needs: Brøderbund and Alliance Call on Software Industry to Design Products That Are Accessible to Customers with Disabilities," 1, box 2, folder 6, Coll. BANC MSS 99/248c, Bancroft Library, University of California, Berkeley.

28. Ibid., 2.

29. Ibid.

30. Ibid.

31. ATA, *Computer Resources for People with Disabilities,* 41.

32. "An Evaluation of Broderbund Software Products: Process Skills, Academic Skill, Access Features, Compatibility with Assistive Technology Devices: For Children with Learning Disabilities and Distinct Learning Styles: A research study conducted jointly by Broderbund Software and The Alliance for Technology Access," 1997, box 2, folder 6, Coll. BANC MSS 99/248c, Bancroft Library, University of California, Berkeley.

33. Ibid.

34. Ibid.

35. ATA, *Computer Resources for People with Disabilities,* 41–42.

36. Bob Glass, "1994 Program Impact Report: Redefining Human Potential: The Partners, Progress and Promise of the Alliance for Technology Access," 19, box 1, folder 5, Coll. BANC MSS 99/248c, Bancroft Library, University of California, Berkeley.

37. ATA, *Computer Resources for People with Disabilities*, 2.

38. Ibid., 8, 127.

39. Ibid., 63.

40. Ibid., 124. Italics original.

41. Ibid., vii.

42. Ibid., viii.

43. Ibid.

44. Ibid., 16–17.

45. Ibid., 117.

46. Ibid., 33.

47. Ibid., 40.

48. ATA, *Quality of Life: Alliance for Technology Access*, 1995, VHS.

49. Ibid.

50. Ibid.

51. Ibid.

52. Ibid.

53. Meeting notes, "Board Retreat (Apr. 26, 1992) Position Statements (as revised by the Executive Committee)," 1992, box 1, folder 3, Coll. BANC MSS 99/185c, Bancroft Library, University of California, Berkeley.

54. *Center for Accessible Technology: A Community-Based Technology Resource Center: 1995 Annual Report*, box 1, folder 2, Coll. BANC MSS 99/185c, Bancroft Library, University of California, Berkeley, 2.

55. Ibid., 3.

56. "1. Overview of Program," InfoUse, accessed Aug. 29, 2012, http://infouse.com /planemath/overview.html.

57. "PlaneMath News Flashes!" InfoUse, accessed Aug. 29, 2012, http://infouse.com /planemath/planemathnews.html.

58. *Center for Accessible Technology: Annual Report 1997*, box 1, folder 2, Coll. BANC MSS 99/185c, Bancroft Library, University of California, Berkeley, 6.

59. Kristen Haugen et al., "Creating Access to PlaneMath," InfoUse, last modified May 15, 1998, accessed Aug. 29, 2012, http://infouse.com/planemath/accessdoc.html.

60. "Plane Math Online Activity," NASA video (2:11), 2001, www.nasa.gov/mov /196829main_066_Plane_Math.mov.

61. Mary Lester, "New Executive Director to Lead Alliance for Technology Access into the 21st Century," ATA press release, Feb. 14, 1997, box 2, folder 6, Coll. BANC MSS 99/248c, Bancroft Library, University of California, Berkeley.

62. Fred Fiedler, "Impressions," *ATAccess*, Alliance for Technology Access, 1997, 2.

63. Jacquelyn Brand, "Parent Advocate for Independent Living, Founder of the Disabled Children's Computer Group and the Alliance for Technology Access," an oral history conducted in 1998–99 by Denise Sherer Jacobson in *Builders and Sustainers of the Independent*

Living Movement in Berkeley, vol. 5, 95, Regional Oral History Office, Bancroft Library, University of California, Berkeley, 2000.

64. Ibid.

65. Ibid., 95.

66. Ibid., 100.

67. "Community Organizations Announce Support for Pacific Telesis—SBC Merger; Company Pledges to Take Leading Role in Universal Service, Create $50 Million Community Technology Fund When Merger Is Complete," The Free Library, Oct. 15, 1996, accessed Aug. 29, 2012, www.thefreelibrary.com/Community Organizations Announce Support for Pacific Telesis—SBC...-a018763350.

68. Bob Tedeschi, "Apple Pulls Plug on Sick Kids' Site," *Wired,* May 15, 1998.

69. Alan Brightman, "Assistive Technology Oral History Project," interview with Chauncy Rucker, Mar. 13, 2008, http://atoralhistory.uconn.edu/podcasts/Brightman.php.

70. Tedeschi, "Apple Pulls Plug on Sick Kids' Site."

71. ATA, "1999/2000 Impact Report: Identity Activities Impact Affiliations," accessed Nov. 25, 2012, 12, http://web.archive.org/web/20100116130313/www.ataccess.org/about/impact2000/default.html. This report has since been removed from the ATA Web site. This copy comes from the Internet Archive capture of the page.

72. Curtis Chong, "Microsoft Takes a Big Step Backward," *Braille Monitor* 40, no. 11 (1997), www.nfb.org/Images/nfb/Publications/bm/bm97/bm971202.htm.

73. Curtis Chong, "Microsoft Promotes Accessibility," *Braille Monitor* 41, no. 5 (1998), www.nfb.org/Images/nfb/Publications/bm/bm98/bm980503.htm.

74. "Remarks by Bill Gates: Microsoft Corporation Accessibility Day," Microsoft, Feb. 19, 1998, accessed Aug. 29, 2012, http://web.archive.org/web/20100909081706/http://www.microsoft.com/presspass/exec/billg/speeches/1998/accessibilityday.aspx. This speech is no longer on Microsoft's Web site; this link is to an Internet Archive capture of the page.

CONCLUSION: **The Promises of Personal Computers**

1. Raymond C. Kurzweil, *The Age of Spiritual Machines: When Computers Exceed Human Intelligence* (New York: Viking, 1999).

2. *Freedom Machines,* directed by Jamie Stobie (Richard Cox Productions, 2004), DVD.

3. Kurzweil, *The Age of Spiritual Machines,* 130.

4. Ibid., 2.

5. Ibid., 135, 234.

6. Ibid., 127.

7. Ibid.

8. Ibid., 128, 150.

9. Ibid., 193.

10. *Freedom Machines.*

11. Ibid.

A Note on Theory, Method, and Sources

This project builds upon a theoretical framework from the history of technology and science and technology studies (STS). Theories in STS relate to the relationship between the use and development of technology, embodied use, and the politics of technology. Using this framework, we can analyze people with disabilities as a specific consumer population group that has a unique and significant relationship to technology. Disability studies also plays an essential role here, including the social theory of disability and the relationship between people with disabilities and assistive technologies. By examining disability through computer technology and civil rights legislation, this history provides a new look at the role technology plays in enacting civil rights and how both interact with the bodies of different users.

Within disability studies, this history follows the development of the social model of disability, which activists and scholars developed in contrast to models of disability such as the medical or individual model. With these earlier models, the problem of disability rested in individuals' bodies, a problem that society could attempt to solve by fixing the body and bringing the person with disabilities closer to normalcy. Within the social model, the problem of disability is caused by barriers keeping people with disabilities from equal rights and equal access to full political lives, and the solution comes from understanding the extent of the limitations placed upon people with disabilities and changing the social environment to accommodate their needs. Disability is connected to the body and to bodily impairments in terms of the assumptions of normalcy which have built up barriers for people with disabilities, but disability is not a medical issue to be treated or fixed; it is a social category and personal identity.

The origin of the social model is explained in Michael Oliver's *Understanding Disability: From Theory to Practice* (New York: St. Martin's Press, 1996). Working within the social model, Lennard J. Davis traces the historical construction of normalcy since the eighteenth century within art, literature, language, and psychology in *Enforcing Normalcy: Disability, Deafness, and the Body* (London: Verso, 1995). Three of the authors I draw upon apply the social model to the history of civil rights for people with disabilities. Claire H. Liachowitz traces the history of disability and legislation during the twentieth century from a social model perspective in *Disability as a Social Construct: Legislative Roots* (Philadelphia: University of Pennsylvania Press, 1988). In his book, *From Good Will to Civil Rights: Transforming Federal Disability Policy* (Philadelphia: Temple University Press, 1984), Richard K. Scotch examines the creation of federal civil rights legislation during the 1970s, particularly Section 504 of the Rehabilitation Act of 1973. He shows the roles played by policy makers and the emerging disability rights movement in enacting the law and its regulations. Joseph P. Shapiro follows a broader look at the same topic in his well-known history of disability rights, *No Pity: People with Disabilities Forging a New Civil Rights Movement* (New York: Three Rivers, 1994). In a contemporary account, Kent Hull explained the legal rights of people with disabilities in the 1970s in *The Rights of Physically Disabled People* (New York: Avon, 1979).

With personal computers, there is a blurring of the line between changing individual bodies to allow for "normal" use versus changing the environment to accommodate different uses. The personal computer is both personal (granting some individuals abilities they otherwise lack) and public (allowing for new spaces of social interaction). Accessible personal computer technologies accommodate uses in both domains. The personal computer is a broad technology that allows for many kinds of uses in different spaces. These technologies, in some instances, offer partial solutions to problems of disability created by social institutions and constructs, particularly regarding communication and the possibility of conducting everyday business from one's home. Technological accommodation has been the political solution to the problem of disability; U.S. legislation attempted to grant people with disabilities equal rights by providing for more and better accessible technologies. The social model seeks to uncover the assumptions behind the unequal position of people with disabilities in society as a way to address the problem at its root and find more inclusive ways to structure society.

Although technology plays an undeniably significant role in the lives of people with disabilities and how they interact with society, disability studies, as a field, has produced relatively few works tackling the relationship between disabled bodies and technology. Scholars who have attempted to study technology while working with disability studies have commented on this lacuna. One place where technology has played a central role as a subject of study is in those works that straddle disability studies and other fields. The intersection of STS and disability studies, in particular, is on the rise, with many developing projects looking at theoretical connections between the two fields and concrete examples in which both fields might talk to each other. In addition, there are a number of histories of assistive technologies, such as prosthetic limbs or cochlear implants, that are used exclusively by people with disabilities.

In his essay, "What Can the Study of Science and Technology Tell Us about Disability?" in *Routledge Handbook of Disability Studies*, ed. Nick Watson, Alan Roulstone, and Carol Thomas (London: Routledge, 2012), 348–59, Stuart Blume examines the hesitation on the part of social model disability studies scholars to treat technology, particularly assistive technologies, as an object of study. He argues that this hesitation comes from the history of assistive technologies originating within rehabilitation sciences, under the medical model of disability. There are a number of histories of assistive technologies by interdisciplinary scholars. David Serlin traces the history of prosthetics in mid-twentieth-century America in his book, *Replaceable You: Engineering the Body in Postwar America* (Chicago: University of Chicago Press, 2004). Interdisciplinary scholars consider the history of prosthetics in a volume coedited by Serlin, Katherine Ott, and Stephen Mihm, *Artificial Parts, Practical Lives: Modern Histories of Prosthetics* (New York: New York University Press, 2002). Though focused on the labor of production instead of the use of an assistive technology, Gregory J. Downey conducts an examination of the history of closed captioning technologies and the workers who caption various types of programming in *Closed Captioning: Subtitling, Stenography, and the Digital Convergence of Text with Television* (Baltimore: Johns Hopkins University Press, 2008).

There are fewer studies of everyday technologies and their relationship with people with disabilities, other than those concerning the removal of architectural barriers preventing the usage of buildings and transportation systems by certain people. Access to buildings is an example of the politics of infrastructure; without the ability to enter and navigate public buildings, people cannot fully participate in social and political life. Computers may not be as essential a part of everyday life as buildings, but during the last few decades, the technology has become ever more integrated into the activities of social life, especially for people with disabilities, and concerns with accessibility are an increasing topic of study.

A few disability studies scholars have considered some of the ways people with disabilities interact with computer technology. In *Enabling Technology: Disabled People, Work, and New Technology* (Philadelphia: Open University Press, 1998), Alan Roulstone discusses the promise new digital technologies offer to change the way work is conducted. Gerard Goggin and Christopher Newell work in between disability studies, media and cultural studies, and STS in order to provide a critical analysis of new media technologies and people with disabilities in *Digital Disability: The Social Construction of Disability in New Media* (Lanham, MD: Rowman and Littlefield, 2003). In *Disability and New Media* (New York: Routledge, 2011), Katie Ellis and Mike Kent build upon the work of Goggin and Newell in examining accessibility and barriers on the Internet today.

Within the history of technology, the historical use of personal computers is mostly absent from academic computer history. Histories of computer development tend to focus on the perspective of computer companies and engineers or blur the division between producers and consumers in considering the role of computer professionals and hobbyists as both early users and developers of personal computer technology. Early computer development resulted in the embedding of certain values into computer technology, such as notions of augmentation. This notion of augmentation is particularly salient for the meanings personal computers hold for people with disabilities, as some early developers saw the computer as functioning like a prosthesis that would develop with its user over time to lead to new abilities.

One of the core texts in the history of computers is Paul Ceruzzi's *A History of Modern Computing*, 2nd ed. (Cambridge, MA: MIT Press, 2003). Ceruzzi chronicles the major developments in computer technologies beginning with the transition from punch-card machines to the first vacuum-tube electronic computers in the late 1940s and early 1950s, to the development of silicon integrated circuits in the late 1960s and early 1970s, the emergence of the personal computer in the late 1970s, and the appearance of networked computers in the 1980s and 1990s. He examines the values of democracy and control that were embedded into the computer throughout this history.

Other significant works within the history of computers analyze more specific parts of their development. In *The Closed World: Computers and the Politics of Discourse in Cold War America* (Cambridge, MA: MIT Press, 1996), Paul Edwards analyzes the computer in its technical form and its importance as a cultural metaphor during the Cold War. He argues that the computer shaped political thought during the Cold War and was, in turn, shaped by it; the politics of Cold War discourse was centered on the interaction of humans and machines. Examining a more recent aspect of computer history, Janet Abbate in *Inventing the Internet* (Cambridge, MA: MIT Press, 1999), studies the creation of pre-Internet networking and the evolution of the military ARPANET into the public World Wide Web. She is interested in the social shaping of Internet technology and draws attention to the role of users in informing the Internet's success.

Examining the history of software from the perspective of computer special-
ists, Nathan Ensmenger analyzes the emergence of programmers as the first users in *The Computer Boys Take Over: Computers, Programmers, and the Politics of Technical Expertise* (Cambridge, MA: MIT Press, 2010). Paul Freiberger and Michael
Swaine consider another group of professionals who influenced the technology
in their popular history examining the origins of Silicon Valley companies, *Fire in
the Valley: The Making of the Personal Computer*, 2nd ed. (New York: McGraw-Hill,
2000).

Some of the ideals and values embedded into the personal computer by early
researchers are discussed in Thierry Bardini's *Bootstrapping: Douglas Engelbart, Co-
evolution, and the Origins of Personal Computing* (Stanford, CA: Stanford University
Press, 2000). Bardini traces Engelbart's pioneering research on developing computers as a means to augment human intelligence and his influence on the development of the personal computer. Christopher Kelty explores the history of free
software and what its creation says about the relationship people desire to have with
information in *Two Bits: The Cultural Significance of Free Software* (Durham, NC:
Duke University Press, 2008). Free software, as a movement, came into being in
the late 1990s; built into it and its software output were values of openness, sharing, and collaboration. Fred Turner discusses the history of the origin of personal
computer values in *From Counterculture to Cyberculture: Stewart Brand, the Whole
Earth Network, and the Rise of Digital Utopianism* (Chicago: University of Chicago
Press, 2006). Turner examines the influence of the 1960s counterculture, particularly Steward Brand's Whole Earth network, on the development of the personal
computer and the change computer technology underwent from being viewed as
industrial and impersonal to a symbol of possible communion and global harmony.

Within the larger history of technology literature, the history of accessible computer technology provides a view of the relationship between society and technology that allows for computers to affect people without removing their agency. Post
technological determinism, technology is neither wholly responsible for changing
society nor completely controllable. Here, the personal computer can and does
change people's lives, particularly in providing new forms of communication and
spaces of social interaction. Users and developers, however, create those technologies and choose how to use them to best fit their needs.

Science and technology studies theories relating to technological use, development, and the relationship of technology to bodies inform the analysis of this
history. The co-construction of users and technology considers the full complexity
of the ways that technology and use affect each other. Computer technology was

developed with certain values embedded in it, which can constrain the choices of users, yet those values became embedded through developers' beliefs about what the technology could do. In turn, while users are affected by design choices, they also choose to use technology in unforeseen ways. The values embedded in the computer are part of the cultural context that tells users what a technology is to be used for and the potential it might have for other uses. This shared understanding of the values built into the technology of the Internet, for example, shapes users' ideas about what can be done with it. The value of the personal computer most relevant to accessibility is that it is a universal technology, capable of being used for any purpose people can imagine, particularly in bringing people together through new forms of communication. Many of its first developers envisioned that the personal computer would be a tool of openness guided by imagination, providing users a means to accomplish any task they could devise, once it had been programmed to do so. This value of universality played into efforts to make the personal computer accessible to people with disabilities, in that the computer should not be limited in who could use it or how they chose to do so.

Albert Borgmann proposes what was, at the time, a new theoretical outlook on the relationship of technology to society, in *Technology and the Character of Contemporary Life* (Chicago: University of Chicago Press, 1984). Borgmann suggests a pluralist theory of technology, in which its complexities are acknowledged, as well as its patterns of development; in the book, he examines the evolving trends of technology and forces of interaction. In another broad work on the relationship between the use and development of technology, *Human-Machine Reconfigurations: Plans and Situated Actions*, 2nd ed. (New York: Cambridge University Press, 2007), Lucy Suchman examines human-machine interaction and the ways people understand new technology. She argues that assumptions regarding the purposeful action of human beings informs the design of interactive machines such as computers but that conceptions of human action as determined by rational plans are inaccurate; this misconception then becomes reflected in the design of machines, which makes the technology less usable.

The conception of use and development from which I work most closely, co-construction, is laid out in Nelly Oudshoorn and Trevor Pinch's edited volume, *How Users Matter: The Co-Construction of Users and Technology* (Cambridge, MA: MIT Press, 2003). These essays complicate our understanding of how users and technology relate to and shape each other, as well as how users are defined and who defines them, by drawing upon previous work in STS, feminist scholarship, and cultural and media studies. Coconstruction is a methodology that examines the full diversity of users, nonusers, and those who speak for users, in every space where technology and users affect each other.

One of the best-known works within STS that blurs the lines between users, spokespeople, and producers is Steven Epstein's *Impure Science: AIDS, Activism, and the Politics of Knowledge* (Berkeley: University of California Press, 1996). In writing about the role AIDS activists played in promoting and testing new drugs, Epstein demonstrates how laypeople can make themselves into experts, coming together within a common identity. Another STS idea I use is that of tinkering, detailed in Karin Knorr-Cetina's *The Manufacture of Knowledge: An Essay on the Constructivist and Contextual Nature of Science* (Oxford: Pergamon, 1981). Tinkering is a way of making a technology fit needs. It could involve scientists making the objects of their research fulfill the goals of their projects or users making a consumer technology work with their needs to solve their problems. In "Making the Pap Smear into the 'Right Tool' for the Job: Cervical Cancer Screening in the USA, circa 1940–95," *Social Studies of Science* 28, no. 2 (1998): 255–90, Monica Casper and Adele Clark demonstrate the necessity of tinkering in making the Pap smear work as the technology of cervical cancer screening.

The promise of more effectively conducting certain tasks is aided by a technology's friendliness, but as the computer became easier to operate, fewer users actually understood the workings of its machinery. The STS concept of "black-boxing" helps unpack this lack of understanding of the inner workings of the computer. In engineering, a "black box" refers to a technology or process where only the inputs and outputs are known; its inner workings or implementation are hidden. STS takes this idea further and considers the social and political implications of black-boxing. Black-boxing makes aspects of a technology unquestionable and, in certain cases from the history of the personal computer, sometimes directly inaccessible. Increasing the black-boxing of the personal computer, in the name of user-friendliness, limited access to its inner machinery, decreasing the flexibility of the computer for alternative uses and preventing access for people with certain kinds of disabilities.

Bruno Latour and Steve Woolgar develop the STS concept of black boxing in *Laboratory Life: The Construction of Scientific Facts* (Princeton, NJ: Princeton University Press, 1986). The design of technology, though not absolutely deterministic of how people might use it, does offer constraints, both physically and in terms of imagined potential. In his article, "Configuring the User: The Case of Usability Trials," in *A Sociology of Monsters: Essays on Power, Technology and Domination*, ed. John Law (London: Routledge, 1991), 57–99, Steve Woolgar uses the metaphor of machine as text in order to explore how users are configured by their computer technology. This idea is not only about how users' actions are delimited by technology design but also about how the technology becomes constructed based on its relationship with its users.

There is a rich group of works within STS and cultural and media studies focused on the embodied use of technology. The authors of these works study the interaction between technologies and bodies in a way that acknowledges that all use is embodied, and they seek to understand the role played by bodily differences. These differences frequently involve issues of gender and race; people with disabilities (and the significant differences between bodies that are revealed when studying disability) are mostly absent. Although these authors discuss assistive technologies, such as prosthetics, they do not, for the most part, analyze the meaning of these technologies in terms of their relationships with disability.

In Anne Marie Balsamo's *Technologies of the Gendered Body: Reading Cyborg Women* (Durham, NC: Duke University Press, 1996), she examines the role technologies play in creating new possibilities for bodies. Also working with ideas of women's bodies and digital technology, Amanda du Preez, in *Gendered Bodies and New Technologies: Rethinking Embodiment in a Cyber-era* (Newcastle upon Tyne: Cambridge Scholars, 2009), argues for the necessity of studying from a perspective of embodiment, as embodiment is a necessary part of being. Within studies of bodies and technology, disability can be an analytic category, as in an essay by Michael L. Dorn, "Beyond Nomadism: The Travel Narratives of a 'Cripple,'" in *Places through the Body*, ed. Heidi J. Nast and Steve Pile (London: Routledge, 2005), 136–52. Criticizing feminist studies' ideas and valuation of the cyborg and nomad as ableist, Dorn examines the writings of Patty Hayes, a disability activist who, over time, came to rethink her relationship with her wheelchair and the built environment in which she attempted to travel and where she encountered barriers.

The development of accessibility in technology crystallized in the 1990s into the idea of universal design, which came to be found in architecture, industrial design, and computer science. Universal design is the result of the history of computer accessibility as well as one of the guiding principles that entrenched accessibility into the development of personal computer technology. Researchers working for the North Carolina State University Center for Universal Design codified the guidelines in 1997; see Bettye Rose Connell, Mike Jones, Ron Mace, Jim Mueller, Abir Mullick, Elaine Ostroff, Jon Sanford, Ed Steinfeld, Molly Story, and Gregg Vanderheiden, "The Principles of Universal Design," North Carolina State University, Center for Universal Design, Raleigh, 1997, www.ncsu.edu/www/ncsu/design/sod5/cud/about_ud/udprinciplestext.htm. They formulated seven principles that would aid developers in designing their products so that as many people as possible could use them without having to further adapt them. These universal design principles have been applied and studied in architecture and design as a way to expand the use of the built environment and consumer products to include the needs of more people. Many of these efforts dealt with explicitly bringing in people with disabilities as users and providing ways to comply with legislative requirements of accessibility.

Universal design changes the understanding of universality by accommodating the needs of all through an awareness of the differences between people. This perspective blurs any distinction between people with disabilities and those without. Disability, then, is a continuum that describes an identity and the ways certain people are prevented from fully participating in society. Universal design is relevant to computer development, as a method to create software and hardware that is usable by as many people as possible. Using a broader version of universal design called universal usability, the computer scientists in the edited volume *Universal Usability: Designing Computer Interfaces for Diverse User Populations* (Chichester, West Sussex: John Wiley and Sons, Ltd, 2007), ed. Jonathan Lazar, lay out methods of creating user interfaces that meet the needs of diverse users. Lazar explains that universal usability imbues universal design with the values of user diversity and greater ease of use. Although universal design did not exist under that name until the late 1990s, its tenets developed throughout the history of personal computer accessibility, increasingly affecting the development process in terms of who the imagined users could be and how their needs were met. Universal design redefined the imagined user not as the universal or average human but as an amalgam of all differences between people. Activists and developers argued for the business benefits of expanding the user base by increasing usability and meeting the needs of people with diverse abilities. There were multiple and sometimes contradictory relationships among usability, accessibility, flexibility, and user-friendliness—where sometimes one quality creates another and sometimes one impedes another.

Regarding primary source materials, much of the history of the Alliance for Technology Access and the Disabled Children's Computer Group came from their archives at the Bancroft Library at the University of California, Berkeley. These archives include annual reports, internal publications, board meeting notes, and a dissertation on the impact of the ATA; see Alliance for Technology Access records, 1987–1999, Coll. BANC MSS 99/248c, Bancroft Library, University of California, Berkeley, and Center for Accessible Technology records, 1985–1998, Coll. BANC

MSS 99/185c Bancroft Library, University of California, Berkeley. The 1985 DCCG annual report and ATAccess newsletter, along with the photographs, are from Jackie Brand's personal collection. A more recent annual report was available at the ATA's Web site until it was removed: Alliance for Technology Access, "1999/2000 Impact Report: Identity Activities Impact Affiliations," accessed November 25, 2012, http://web.archive.org/web/20100116130313/http://www.ataccess.org/about /impact2000/default.html.

The ATA's book, *Computer Resources for People with Disabilities: A Guide to Exploring Today's Assistive Technology* (Alameda, CA: Hunter House Inc., 1994), was a source for much of the group's philosophy toward technology for people with disabilities. Other important publications from the two groups included Jacquelyn Brand, "Families Working Together," *Exceptional Parent,* October 1985, 17–18, and "Developing a Parent/Community Technology Resource Center," *Closing the Gap,* April 12, 1986, 1. Details on various ATA and DCCG projects have been covered in a number of news articles and press releases. The PlaneMath project still exists as a Web site, available at http://in fouse.com/planemath. NASA's Web site also contains some information on PlaneMath, including a video talking about the project; see www.nasa.gov /mov/196829main_066_Plane_Math.mov.

For personal information on the Brand family, I drew upon two oral histories that Jackie Brand participated in: "Parent Advocate for Independent Living, Founder of the Disabled Children's Computer Group and the Alliance for Technology Access," an oral history conducted in 1998–99 by Denise Sherer Jacobson in *Builders and Sustainers of the Independent Living Movement in Berkeley,* vol. 5, Regional Oral History Office, Bancroft Library, University of California, Berkeley, 2000, and "Assistive Technology Oral History Project," interview with Chauncy Rucker, November 1, 2007, http://atoralhistory.uconn.edu/podcasts/brand.php.

Apple Computer's main archive at Stanford University does not contain any materials on the company's accessibility work. Most of my information on Apple came from the ATA and DCCG materials or assorted publications and news articles. Alan Brightman took part in an interview that provided much of the history of his work at the Apple Office of Special Education and Rehabilitation, "Assistive Technology Oral History Project," interview with Chauncy Rucker, March 13, 2008, http://atoral history.uconn.edu/podcasts/Brightman.php. For general Apple history, I used Steve Wozniak's autobiography, cowritten with Gina Smith, *iWoz: Computer Geek to Cult Icon: How I Invented the Personal Computer, Co-founded Apple, and Had Fun Doing It* (New York: W. W. Norton and Co, 2006), as well as two nonacademic histories, Steven Levy's *Insanely Great: The Life and Times of Macintosh, the Computer That Changed Everything* (New York: Penguin, 1994) and Owen W. Linzmayer's *Apple Confidential 2.0: The Definitive History of the World's Most Colorful Company* (San Francisco: No Starch Press, 2004).

Compared to Apple, IBM saved more materials on its accessibility efforts. IBM's corporate archive provided me with a number of articles from the company's *Think* magazine on its employment of people with disabilities and the technological accommodations for them. IBM also published a booklet on the groups within the company that conducted accessibility work, IBM National Support Center for Persons with Disabilities, *Technology for Persons with Disabilities: An Introduction* (n.p.: IBM, 1990). In addition, IBM has archived materials digitally on its Web site, including user manuals for software applications and a chronology of work for people with disabilities. The history of IBM is also chronicled by Emerson Pugh in *Building IBM: Shaping an Industry and Its Technology* (Cambridge, MA: MIT Press, 1995).

In discussing civil rights legislation, I drew upon the actual laws themselves (the Architectural Barriers Act of 1968, the Rehabilitation Act of 1973, the Education for All Handicapped Children Act of 1975, the Technology-Related Assistance for Individuals with Disabilities Act of 1988, and the Americans with Disabilities Act of 1990), their regulations, and testimony submitted prior to their passage. I also used publications from the National Council on the Handicapped and its later incarnation, the National Council on Disability. Gallaudet University maintains a digital archive of materials related to the Deaf President Now protest, including flyers, position statements from protesters, and letters of support. These can be found at www.gallaudet.edu/Gallaudet_University/About_Gallaudet/DPN_Home /Issues.html.

The history of the ACM Special Interest Group on Computing and the Physically Handicapped comes from articles published in the *SIGCAPH Newsletter*, active since 1974. The newsletter included articles on research and technology for people with disabilities as well as discussions of internal, organizational issues. Information on the Johns Hopkins contest on Applications of Personal Computing to Aid the Handicapped was published in various newsletters and magazines, including the *SIGCAPH Newsletter*. The submissions for all finalists were collected into a book, *Proceedings of the Johns Hopkins First National Search for Applications of Personal Computing to Aid the Handicapped* (Los Angeles: IEEE Computer Society, 1981). The *SIGCAPH Newsletter* also provided material on Murray Turoff's computerized conferencing research, although I drew the bulk of this history from

Turoff and Starr Roxanne Hiltz's book, *The Network Nation: Human Communication via Computer* (Reading, MA: Addison-Wesley, 1978).

Details on accessibility pertaining to the Graphical User Interface and operating systems came primarily from publications from the Trace Center at the University of Wisconsin–Madison and the National Federation of the Blind's digital archive of its *Braille Monitor* newsletter, maintained at https://nfb.org/braille-monitor#Online Access. Other Trace Center publications provided an early look at how researchers and developers were considering personal computer accessibility.

Other essential primary historical materials included Frank G. Bowe's *Personal Computers and Special Needs* (Berkeley, CA: Sybex, 1984) and Peter A. McWilliams's *Personal Computers and the Disabled* (Garden City, NY: Garden Press, 1984). Both of these books were guides to buying computer technology for users with disabilities. Joseph J. Lazzaro's *Adapting PCs for Disabilities* (Reading, MA: Addison-Wesley, 1996) covers similar topics a decade later. Focusing specifically on the use of personal computer technologies in special education classrooms is Dolores Hagan's *Microcomputer Resource Book for Special Education* (Reston, VA: Reston Pub. Co., 1984).

Work done by blind computer programmers in the 1960s is described in the Committee on Professional Activities of the Blind's *The Selection, Training, and Placement of Blind Computer Programmers* (n.p.: Association for Computing Machinery, 1966) and Theodor D. Sterling, M. Lichstein, F. Scarpino, D. Stuebing, and William Stuebing, "Professional Computer Work for the Blind," *Communications of the ACM* 7, no. 4 (1964): 228–51.

The primary sources that I compare in my conclusion are Raymond C. Kurzweil's *The Age of Spiritual Machines: When Computers Exceed Human Intelligence* (New York: Viking, 1999) and the documentary *Freedom Machines,* directed by Jamie Stobie (Richard Cox Productions, 2004), DVD. In Kurzweil's book, he lays out his philosophy toward technology and human bodies, as well as his predictions for how future technology would change what it means to be human. *Freedom Machines* provides both a practical and an emotional look at the necessity of accessible technologies for people with disabilities and the difficulties they have encountered in trying to learn about and acquire such technology.

Index